ALBERTA

Ian Sheldon & Tamara Eder

First printed in 2001 10 9 8 7 6 5 4 3
Printed in Canada

THE PUBLISHER: LONE PINE PUBLISHING

1808 B Street NW Suite 140
Auburn, WA 98001
USA

10145-81 Avenue
Edmonton, AB T6E 1W9
Canada

Lone Pine Publishing Website: http://www.lonepinepublishing.com

Canadian Cataloguing in Publication Data

Sheldon, Ian, (date)
Animal tracks of Alberta

Includes bibliographical references and index.
ISBN 1-55105-312-8

1. Animal tracks—Alberta—Identification. I. Eder, Tamara, (date) II. Title.
QL768.S521 2001 591.47'9 C00-911537-4

Editorial Director: Nancy Foulds
Editor: Volker Bodegom
Proofreader: Randy Williams
Production Manager: Jody Reekie
Design, layout and production: Volker Bodegom, Monica Triska
Cover Design: Robert Weidemann
Cartography: Volker Bodegom
Animal illustrations: Gary Ross, Horst Krause, Ian Sheldon, Ewa Pluciennik
Track illustrations: Ian Sheldon
Cover illustration: Mule Deer by Gary Ross

We acknowledge the financial support of the Government of Canada through the Book Publishing Industry Development Program (BPIDP) for our publishing activities.

PC: P1

CONTENTS

INTRODUCTION

If you have ever spent time with an experienced tracker, or perhaps a veteran hunter, then you know just how much there is to learn about the subject of tracking and just how exciting the challenge of tracking animals can be. Maybe you think that tracking is no fun, because all you get to see is the animal's prints. What about the animal itself—is that not much more exciting? Well, for most of us who don't spend a great deal of time in the beautiful wilderness of Alberta, the chances of seeing the majestic Moose or the fun-loving River Otter are slim. The closest that we may ever get to some animals will be through their tracks, and they can inspire a very intimate experience. Remember, you are following in the footsteps of the unseen—animals that are in pursuit of prey, or perhaps being pursued as prey.

This book offers an introduction to the complex world of tracking animals. Sometimes tracking is easy. At other times it is an incredible challenge that leaves you wondering just what animal made those unusual tracks. Take this book into the field with you, and it can provide some help with the first steps to identification. Animal tracks and trails are this book's focus; you will learn to recognize subtle differences for both. There are, of course, many additional signs to consider, such as scat and food caches, all of which help you to understand the animal that you are tracking.

Remember, it takes many years to become an expert tracker. Tracking is one of those skills that grows with you as you acquire new knowledge in new situations. Most importantly, you will have an intimate experience with nature. You will learn the secrets of the seldom-seen. The more you discover, the more you will want to know. And, by developing a good understanding of tracking, you will gain an excellent appreciation of the intricacies and delights of our marvellous natural world.

How to Use This Book

Most importantly, take this book into the field with you! Relying on your memory is not an adequate way to identify tracks. Track identification has to be done in the field, or with detailed sketches and notes that you can take home.

Coyote

Much of the process of identification involves circumstantial evidence, so you will have much more success when standing beside the track.

This book is laid out so as to be easy to use. Beginning on p. 152, there is a quick reference appendix to the tracks of all the animals illustrated in the book. This appendix is a fast way to familiarize yourself with certain tracks and the content of the book, and it guides you to the more informative descriptions of each animal and its tracks.

Each animal's description is illustrated with the appropriate footprints and the track patterns that it usually leaves. Although these illustrations are not exhaustive, they do show the tracks or groups of prints that you will most likely see. You will find a list of dimensions for the tracks, giving the general range, but there will always be extremes, just as there are with people who have unusually small or large feet. Under the category 'Size' (of animal), the 'greater-than' sign (>) is used when the size difference between the sexes is pronounced.

If you think that you may have identified a track, check the 'Similar Species' section for that animal. This section is designed to help you confirm your conclusions by pointing out other animals that leave similar tracks and showing you ways to distinguish among them.

As you read this book, you will notice an abundance of words such as 'often,' 'mostly' and 'usually.' Unfortunately, tracking will never be an exact science; we cannot expect animals to conform to our expectations, so be prepared for the unpredictable.

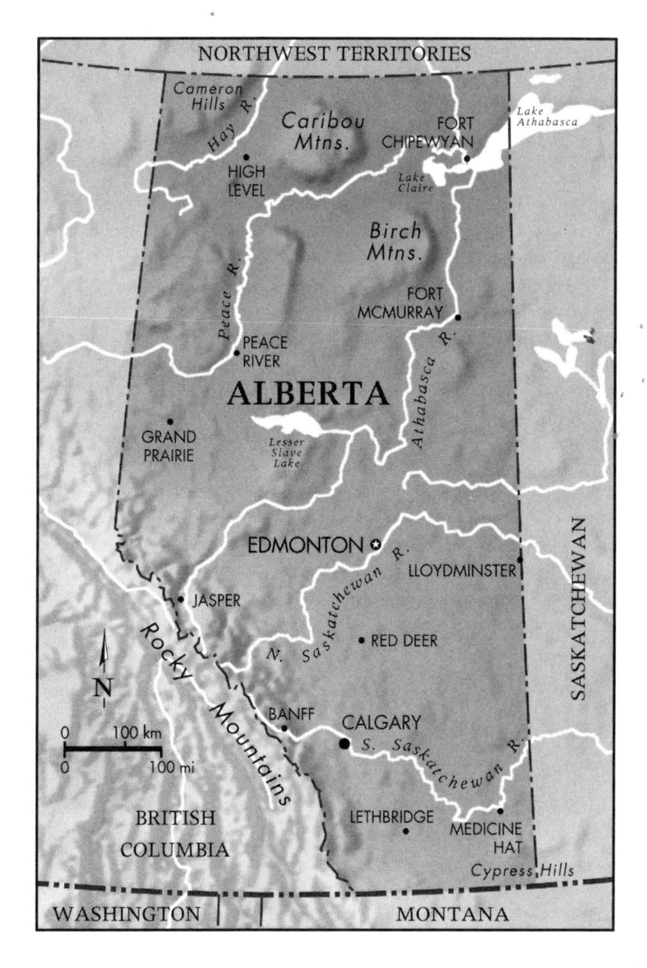
NORTHWEST TERRITORIES
Cameron Hills
Hay R.
Caribou Mtns.
FORT CHIPEWYAN
Lake Athabasca
HIGH LEVEL
Lake Claire
Birch Mtns.
Peace R.
FORT MCMURRAY
PEACE RIVER
Athabasca R.
ALBERTA
GRAND PRAIRIE
Lesser Slave Lake
EDMONTON
LLOYDMINSTER
N. Saskatchewan R.
JASPER
RED DEER
SASKATCHEWAN
N
Rocky Mountains
BANFF
CALGARY
S. Saskatchewan R.
0
100 km
0
100 mi
BRITISH COLUMBIA
LETHBRIDGE
MEDICINE HAT
Cypress Hills
WASHINGTON
MONTANA

Tips on Tracking

As you flip through this guide, you will notice clear, well-formed prints. Do not be deceived! It is a rare track that will ever show so clearly. For a good, clear print, the perfect conditions are slightly wet, shallow snow that isn't melting, or slightly soft mud that isn't actually wet. These conditions can be rare—most often you will be dealing with incomplete or faint prints, where you cannot even be sure of the number of toes.

Should you find yourself looking at a clear print, then the job of identification is much easier. There are a number of key features to look for: Measure the length and width of the print, count the number of toes, check for claw marks and note how far away they are from the body of the print, and look for a heel mark. Keep in mind more subtle features, such as the spacing between the toes and whether or not they are parallel, and whether fur on the sole of the foot has made the print less clear.

When you are faced with the challenge of identifying an unclear print—or even if you think that you have made a successful identification from one print alone—look beyond the single footprint and search out others. Do not rely on the dimensions of just one print, but collect measurements from several prints to get an average impression. Even the prints within one trail can show a lot of variation.

Try to determine which is the fore print and which is the hind, and remember that many animals are built very differently from humans, having larger forefeet than hind feet. Sometimes the prints will overlap, or they can be directly on top of one another in a direct register. For some animals, the fore and hind prints are pretty much the same.

Check out the pattern that the tracks make together in the trail and follow the trail for as many paces as is necessary for you to become familiar with the pattern. Patterns are very important and can be the distinguishing feature between different animals with otherwise similar tracks.

Follow the trail for some distance, because it can give you some vital clues. For example, the trail may lead you to a tree, indicating that the animal is a climber—or it may lead down into a burrow. This part of tracking can be the most rewarding, because you are following the life of the animal as it hunts, runs, walks, jumps, feeds or tries to escape a predator.

Take into consideration the habitat. Sometimes habitat alone will allow you to distinguish very similar tracks—one species might be found on riverbanks, whereas another might be encountered just in dense forest.

Think about your geographical location, too, because some animals have a limited range. This consideration can rule out some species and help you with your identification.

Remember that every animal will at some point leave a print or trail that looks just like the print or trail of a completely different animal! Finally, keep in mind that if you track quietly, you might catch up with the maker of the prints.

Terms & Measurements

Some of the terms used in tracking can be rather confusing, and they often depend on personal interpretation. For example, what comes to your mind if you see the word 'hopping'? Perhaps you see a person hopping about on one leg—or perhaps you see a rabbit hopping through the countryside. Clearly, one person's perception of motion can be very different from another's. Some useful terms are explained on the next few pages, to clarify what is meant in this book, and, where appropriate, how the measurements given fit in with each term.

The following terms are sometimes used loosely and interchangeably—for example, a rabbit might be described as 'a hopper' and a squirrel as 'a bounder,' yet both leave the same pattern of prints in the same sequence.

Ambling: Fast, rolling walking.

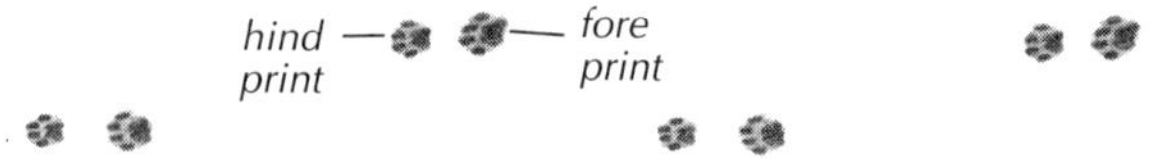

hind print
fore print

Bounding: A gait of four-legged animals in which the two hind feet land simultaneously, usually registering in front of the fore prints. It is common in rodents and the rabbit family. 'Hopping' or 'jumping' can often be substituted.

hind prints
fore prints

Gait: An animal's gait describes how it is moving at some point in time. Different gaits result in different observable trail characteristics.

Galloping: A gait used by animals with four even-length legs, such as dogs, moving at high speed, hind feet registering in front of forefeet.

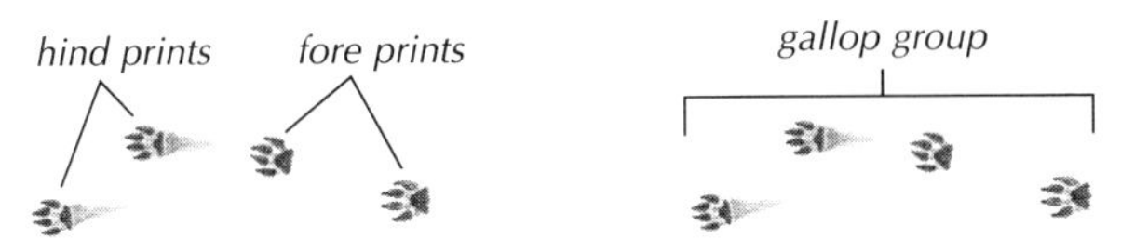

Hopping: Similar to bounding. With four-legged animals, it is usually indicated by tight clusters of prints, fore prints set between and behind the hind prints. A bird hopping on two feet creates a series of paired tracks along its trail.

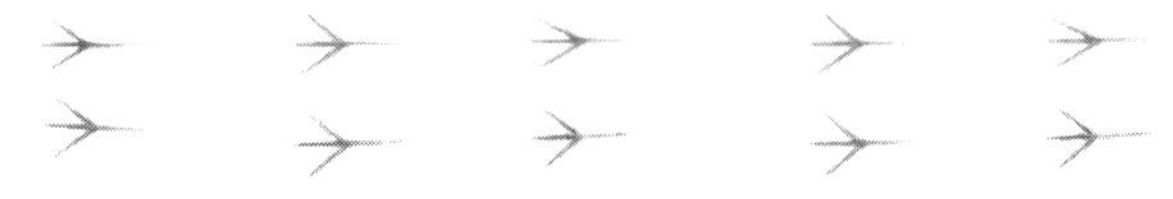

Loping: Like galloping but slower, with each foot falling independently and leaving a trail pattern that consists of groups of tracks in the sequence fore-hind-fore-hind, usually roughly in a line.

Mustelids (weasel family) often use ***2×2 loping***, in which the hind feet register directly on the fore prints. The resulting pattern has angled, paired tracks.

Running: Like galloping, but applied generally to animals moving at high speed. Also used for two-legged animals.

Stotting (applies to the Mule Deer only): Describes the action of taking off from the ground and landing on all four feet at once, in pogo-stick fashion.

Trotting: Faster than walking, slower than running. The diagonally opposite limbs move simultaneously; that is, the right forefoot with the left hind, then the left forefoot with the right hind. This gait is the natural one for canids (dog family), short-tailed shrews and voles.

Canids may use ***side-trotting***, a fast trotting in which the hind end of the animal shifts to one side. The resulting track pattern has paired tracks, with all the fore prints on one side and all the hind prints on the other.

Walking: A slow gait in which each foot moves independently of the others, resulting in an alternating track pattern. This gait is common for felines (cat family) and deer, as well as wide-bodied animals, such as bears and porcupines. The term is also used for two-legged animals.

Other Tracking Terms:

Dewclaws: Two small, toe-like structures set above and behind the main foot of most hoofed animals.

Direct Register: The hind foot falls directly on the fore print.

Double Register: The hind foot registers so as to overlap the fore print only slightly or falls beside it, so that both prints can be seen at least in part.

Dragline: A line left in snow or mud by a foot or the tail dragging over the surface.

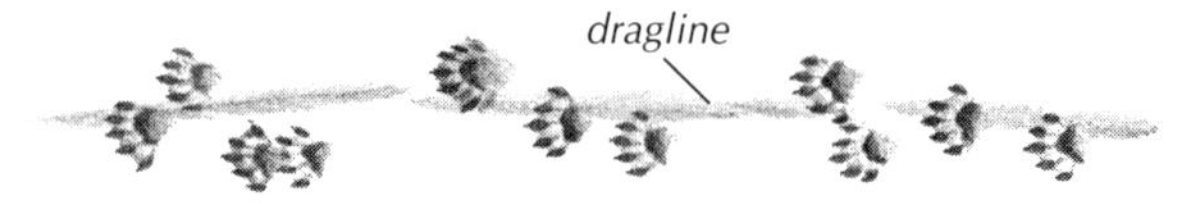

Gallop Group: A track pattern of four prints made at a gallop, usually with the hind feet registering in front of the forefeet (see '**galloping**' for illustration).

Height: Taken at the animal's shoulder.

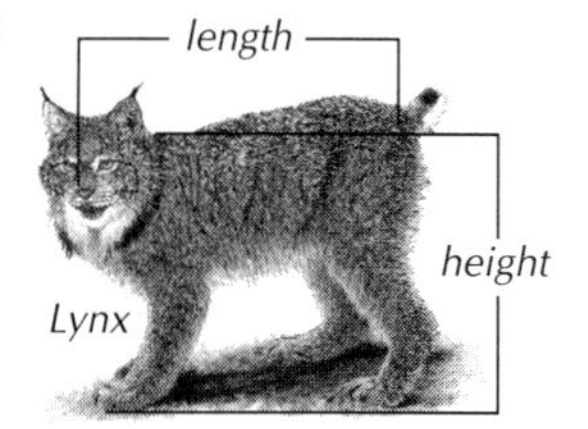

Length: The animal's body length from head to rump, not including the tail, unless otherwise indicated.

Metacarpal Pad: A small pad near the palm pad or between the palm pad and heel on the forefeet of bears and members of the weasel family.

Print (also called '***track***'): Fore and hind prints are treated individually. Print dimensions given are 'length' (including claws—maximum values may represent occasional heel register for some animals) and 'width.' A group of prints made by each of the animal's feet makes up a track pattern.

Register: To leave a mark—said about a foot, claw or other part of an animal's body.

Retractable: Describes claws that can be pulled in to keep them sharp, as with the cat family; these claws do not register in the prints. Foxes have semi-retractable claws.

Sitzmark: The mark left on the ground by an animal falling or jumping from a tree.

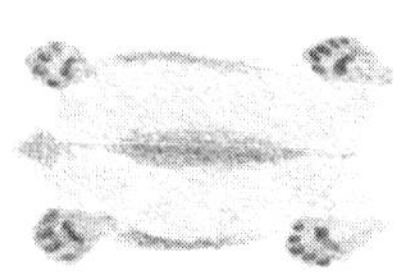

Straddle: The total width of the trail, all prints considered.

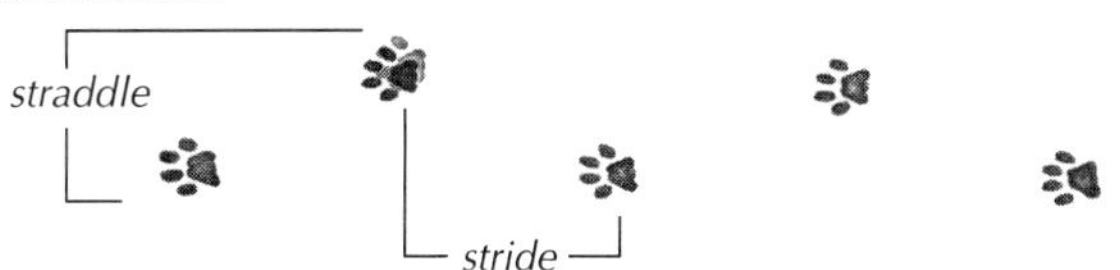

Stride: For consistency among different animals, the stride is taken as the distance from the centre of one print (or print group) to the centre of the next one. Some books may use the term 'pace.'

Track: Same as '***print***.'

Track Pattern: The pattern left after each foot registers once; a set of prints, such as a gallop group.

Trail: A series of track patterns; think of it as the path of the animal.

MAMMALS

River Otter

Bison

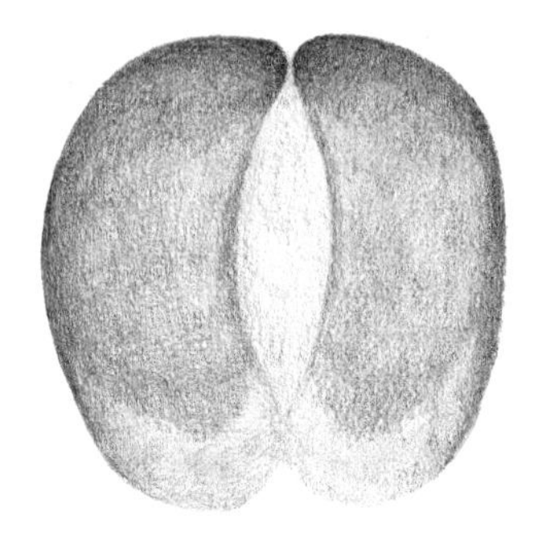

Fore and Hind Prints
Length: (10–15 cm) 4–6 in
Width: (10–15 cm) 4–6 in

Straddle
25–55 cm (10–21 in)

Stride
Walking: 35–80 cm (14–32 in)

Size (bull>cow)
Height: 1.5–1.8 m (5–6 ft)
Length: 3–3.7 m (10–12 ft)

Weight
Male: 360–900 kg (800–2000 lb)
Female: 320–500 kg (700–1100 lb)

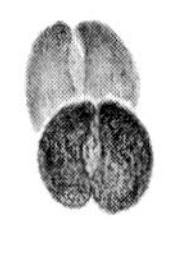

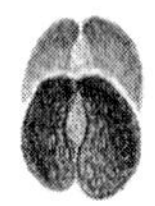

walking

BISON (Buffalo)

Bison bison

An estimated 70 million Bison once roamed North America. As few as 1500 survived the wholesale Bison slaughter of the nineteenth century. Today, as a result of a major effort to save this magnificent beast from extinction, over 350,000 Bison inhabit scattered protected areas and ranches across the continent.

In a Bison's alternating walking pattern, the slightly smaller hind foot usually registers on or near the fore print. On firm ground, only the outer edge of the hoof may register, but in soft mud or snow the whole foot registers—perhaps the dewclaws, too. Foot drag is common. The abundant 'pies' may be mistaken for those of domestic cattle. Additional signs of Bison include rubbing posts or trees that have tufts of distinctive brown hair hanging from them, and the large pits in which Bison wallow. Do not be fooled by a Bison's calm exterior—a hefty bull Bison can inflict serious injury!

Similar Species: Domestic Cattle (*Bos* spp.) prints are similar. On firm surfaces, Horse (p. 36) prints can resemble Bison prints.

Caribou

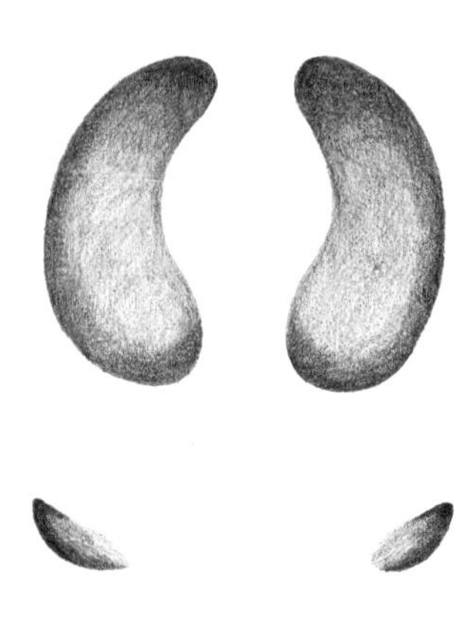

Fore and Hind Prints
Length: 7.5–12 cm (3–4.7 in)
Length with dewclaws:
to 20 cm (8 in)
Width: 10–15 cm (4–5.8 in)

Straddle
23–35 cm (9–14 in)

Stride
Walking: 40–80 cm (16–32 in)
Running: to 1.5 m (5 ft)
Group length: to 2.7 m (9 ft)

Size (buck>doe)
Height: 1.1–1.2 m (3.5–4 ft)
Length: 2–2.6 m (6.5–8.5 ft)

Weight
70–270 kg (150–600 lb)

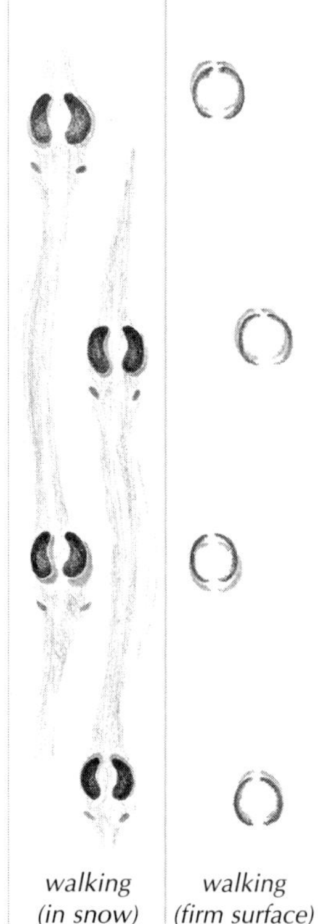

walking (in snow)

walking (firm surface)

CARIBOU

Rangifer tarandus

There are only a few places in Alberta where you might be privileged to see the elegant antlers of the male Caribou. The Caribou is a true wilderness animal, and its sensitivity to human encroachment has reduced its range to protected parks and remote areas of the north, where it prefers to feed in groups in open country.

The soft inner sole of the Caribou's hoof hardens and shrinks in winter, leaving a firm outer wall that makes neat circles on firmer surfaces. The snowshoe-like hoof spreads wide, making a distinctive large, rounded print. Big dewclaws, which help the Caribou to distribute its weight on snow, register behind the forefeet, but rarely the hind feet; the dewclaws move sideward as the animal increases speed. Foot drag in snow is common; examine the tracks to see how the Caribou swings its legs as it walks. Also, look for for scrape marks where Caribou have dug for lichens hidden beneath the snow.

Similar Species: The shape and size of Caribou prints minimize confusion. Moose (p. 22) prints are larger and less round, with a wider straddle. Horse (p. 36) prints are larger and not in two parts.

Moose

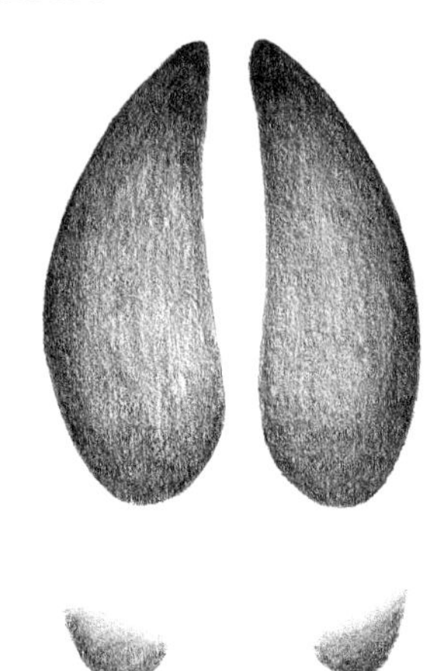

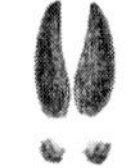

walking

Fore and Hind Prints
Length: 10–18 cm (4–7 in)
Length with dewclaws: to 28 cm (11 in)
Width: 9–15 cm (3.5–6 in)

Straddle
22–50 cm (8.5–20 in)

Stride
Walking: 45–90 cm (1.5–3 ft)
Trotting: to 1.2 m (4 ft)

Size (bull>cow)
Height: 1.5–2 m (5–6.5 ft)
Length: 2.1–2.6 m (7–8.5 ft)

Weight
270–500 kg (600–1100 lb)

MOOSE

Alces alces

The impressive male Moose, the largest of the deer, has a massive rack of antlers. Moose are usually solitary, though you may see a cow with her calf. Despite its placid appearance, a Moose may charge humans if approached.

Though ungainly in shape, the Moose moves gracefully, leaving a neat alternating walking pattern. The hind feet direct or double register on the fore prints. Long legs allow for easy movement in snow. Dewclaws—which give extra support for the animal's great weight—register in prints more than 3 cm (1.2 in) deep, but far behind the hoof. In summer, look for tracks in mud beside ponds and other wet areas, where Moose especially like to feed; they are excellent swimmers. In winter, Moose feed in willow flats and coniferous forests, leaving a distinct browseline (highline). Ripped stems and scraped bark, 1.8 m (6 ft) or more above the ground, are additional signs of Moose.

Similar Species: Caribou (p. 20), Elk (p. 24) and deer (pp. 26–29) tracks are all smaller but may be mistaken for juvenile Moose tracks.

Elk

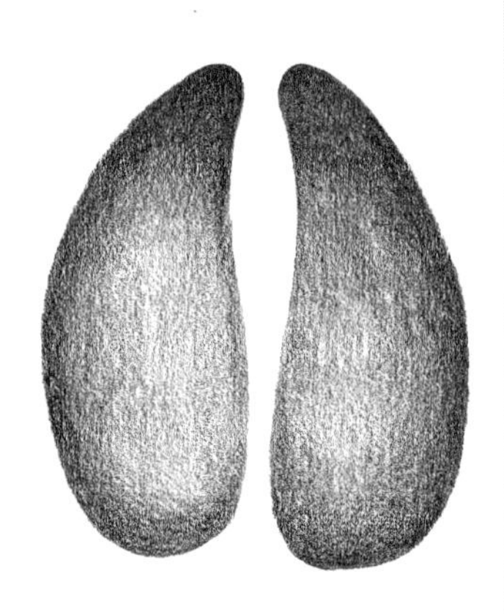

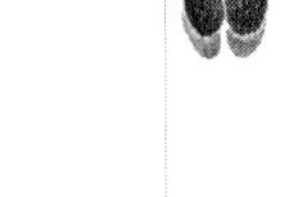

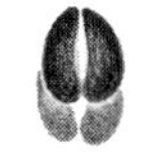

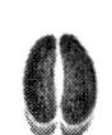

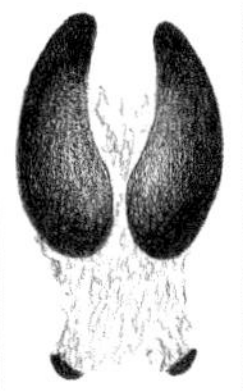

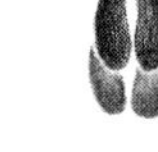

Fore and Hind Prints
Length: 8–13 cm (3.2–5 in)
Width: 6.5–11 cm (2.5–4.5 in)

Straddle
18–30 cm (7–12 in)

Stride
Walking: 40–85 cm (16–34 in)
Galloping: 1–2.4 m (3.3–8 ft)
Group length: to 1.9 m (6.3 ft)

Size (bull>cow)
Height: 1.2–1.5 m (4–5 ft)
Length: 2–3 m (6.5–10 ft)

Weight
230–450 kg (500–1000 lb)

gallop print

walking

ELK (Wapiti)

Cervus elaphus

Common in the meadows and open forests of the mountains and foothills, the Elk is almost gone from the prairies; the population in Elk Island National Park is one of the few that remains. Female Elk and young are often seen in social herds. They like to feed in forest openings and meadows. Stags, who prefer to go solo, are easily recognized by their magnificent racks of antlers and their distinctive bugling in late August. Look for Elk tracks in the soft mud beside summer ponds, where Elk like to drink and sometimes splash around.

The Elk leaves a neat alternating walking pattern of large, rounded prints, often in well-worn winter paths. The hind foot will sometimes double register slightly in front of the fore print. In deeper snow, or if an Elk gallops (with its toes spread wide), the dewclaws may register.

Similar Species: Deer (pp. 26–29) prints can be similar but are generally smaller. Moose (p. 22) prints are similar but larger.

Mule Deer

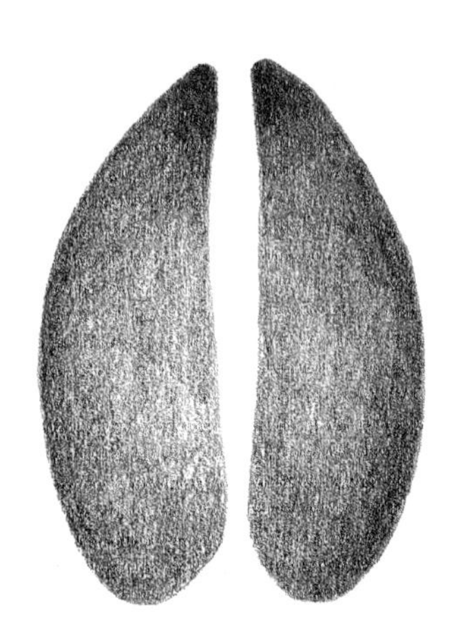

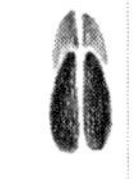

walking *stot group*

Fore and Hind Prints
Length: 5–8.5 cm (2–3.3 in)
Width: 4–6.5 cm (1.6–2.5 in)

Straddle
13–25 cm (5–10 in)

Stride
Walking: 25–60 cm (10–24 in)
Jumping: 2.7–5.8 m (9–19 ft)

Size (buck>doe)
Height: 90–110 cm (3–3.5 ft)
Length: 1.2–2 m (4–6.5 ft)

Weight
45–200 kg (100–450 lb)

MULE DEER
(Black-tailed Deer)

Odocoileus hemionus

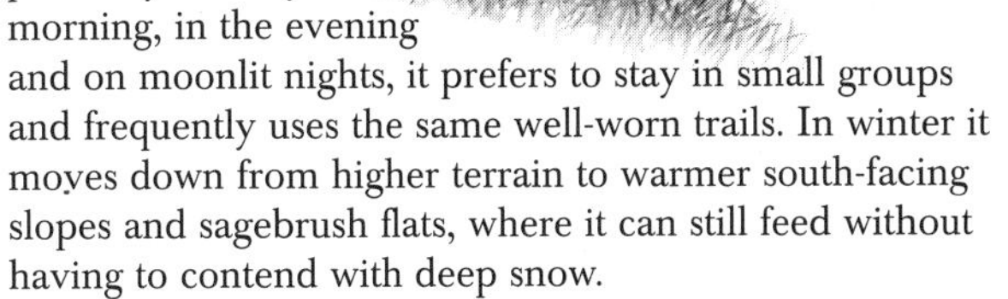

The widespread Mule Deer frequents meadows, open woodlands and plains. Active primarily in early morning, in the evening and on moonlit nights, it prefers to stay in small groups and frequently uses the same well-worn trails. In winter it moves down from higher terrain to warmer south-facing slopes and sagebrush flats, where it can still feed without having to contend with deep snow.

The Mule Deer makes sharply pointed heart-shaped prints. In deep mud or snow, or when the animal is moving quickly, the dewclaws register, closer to the hoof on the fore prints than on the hind ones. The neat alternating walking pattern shows the hind print on top of the fore print. At high speed this deer has a unique gait—stotting—in which it jumps with all its feet leaving or striking the ground at once, and the toes splay to distribute the animal's weight and give better footing.

Similar Species: The White-tailed Deer (p. 28), with near-identical prints, prefers denser cover and has a different high-speed track pattern with a shorter stride. Elk (p. 24) prints are longer and wider. Pronghorn Antelope (p. 30) prints have wider bases.

White-tailed Deer

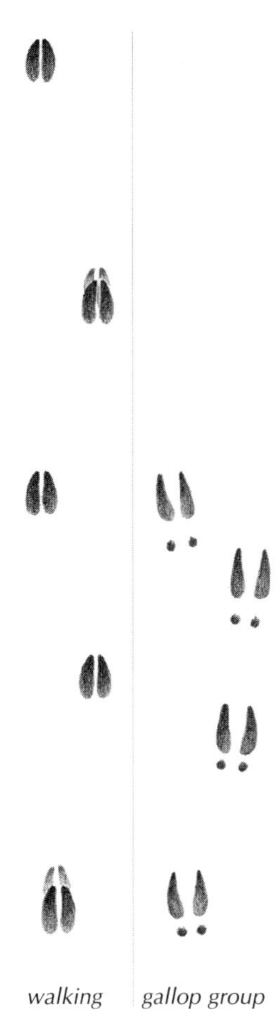

Fore and Hind Prints
Length: 5–9 cm (2–3.5 in)
Width: 4–6.5 cm (1.6–2.5 in)

Straddle
13–25 cm (5–10 in)

Stride
Walking: 25–50 cm (10–20 in)
Galloping: 1.8–4.5 m (6–15 ft)

Size (buck>doe)
Height: 90–110 cm (3–3.5 ft)
Length: to 1.9 m (6.3 ft)

Weight
55–160 kg (120–350 lb)

WHITE-TAILED DEER

Odocoileus virginianus

The keen hearing of this deer guarantees that it knows about you before you know about it. Frequently, all that we see is its conspicuous white tail in the distance as it gallops away, earning this deer the nickname 'Flagtail.' The adaptable White-tailed Deer can be found in small groups at the edges of forests and in brushlands. Deer can be common sights around ranches and in residential areas.

This deer's prints are heart-shaped and pointed. Its alternating walking track pattern shows the hind prints direct registered or double registered on the fore prints. In snow, or when a deer gallops on soft surfaces, the dewclaws register. This flighty deer gallops in the usual style, leaving hind prints ahead of fore prints, with toes spread wide for steadier, safer footing.

Similar Species: The Mule Deer (p. 26), with near-identical tracks, prefers more-open terrain, and at high speed it stots instead of gallops. Elk (p. 24) prints are longer and wider. Juvenile Moose (p. 22) tracks may be confused with large deer tracks. Antelopes (p. 30) prefer open spaces, and their prints have wider bases.

Pronghorn Antelope

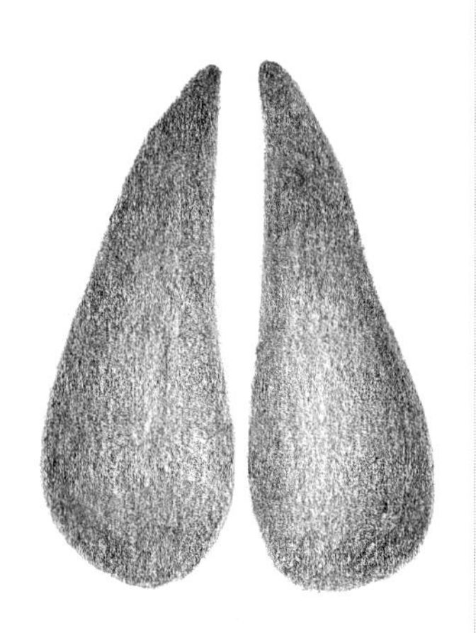

Fore and Hind Prints
Length: 8.5 cm (3.3 in)
Width: 6.5 cm (2.5 in)

Straddle
9–23 cm (3.5–9 in)

Stride
Walking: 20–48 cm (8–19 in)
Galloping: 4.3 m (14 ft) or more

Size (buck>doe)
Height: 90 cm (3 ft)
Length: 1.2–1.5 m (3.8–5 ft)

Weight
75–130 lb (34–60 kg)

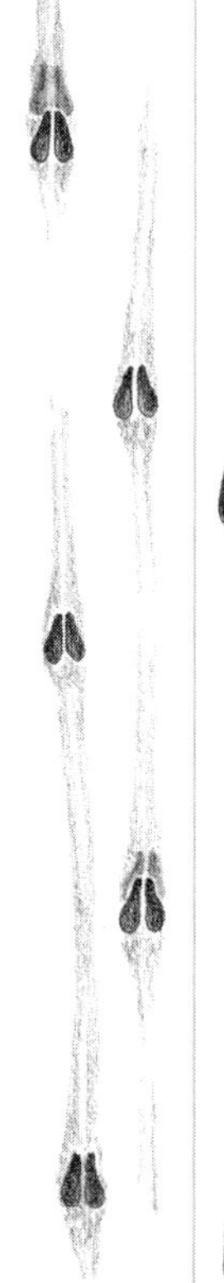

walking

gallop group

PRONGHORN ANTELOPE

Antilocapra americana

The graceful Pronghorn Antelope frequents the wide-open grasslands and sagebrush plains of southern Alberta. Pronghorns gather in groups of up to one dozen animals in summer and as many as one hundred in winter, when they prefer to feed in areas where the snow has been blown away. Unlike deer, the Pronghorn Antelope—one of North America's fastest animals—runs for fun, easily attaining constant speeds of 65 km/h (40 mph) and short spurts of as much as 95 km/h (60 mph).

A Pronghorn print has a pointed tip and a broad base. This animal does not have dewclaws. The hind prints usually register directly on top of the fore prints, making a tidy alternating track. During its frequent gallops, the toe tips splay wide. The faster the antelope moves, the greater the distance between gallop groups. Pronghorns tend to drag their feet in snow.

Similar Species: Deer prints (pp. 26–29) show dewclaws, narrower toe bases and shorter strides between gallop groups.

Mountain Goat

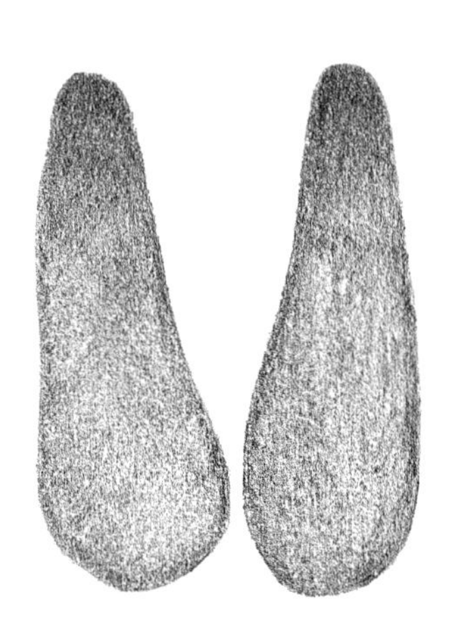

walking

Fore and Hind Prints
Length: 6.5–9 cm (2.5–3.5 in)
Width: 5–8.5 cm (2–3.3 in)

Straddle
17–30 cm (6.5–12 in)

Stride
Walking: 25–48 cm (10–19 in)

Size (billy>nanny)
Height: 90–110 cm (3–3.5 ft)
Length: 1.5–1.8 m (5–6 ft)

Weight
45–140 kg (100–300 lb)

MOUNTAIN GOAT

Oreamnos americanus

Spotting the dazzling white coat of a Mountain Goat is truly a high-mountain wilderness experience. This goat has a preference for high terrain, usually on rocky slopes above the treeline. The Mountain Goat is rarely seen at close range; its tracks in the snow may be your best clue that it is around.

The goat's squarish print shows long, widely spreading toes. The hard rim and soft middle of the foot help this agile goat clamber over the most unlikely crags at remarkable speeds, but even the Mountain Goat can make a fatal mistake! In deeper snow the feet may leave draglines, and the dewclaws may register. The alternating walking pattern shows a double register, hind over fore.

Similar Species: Deer (pp. 26–29) or Bighorn Sheep (p. 34) prints are smaller, narrower and more pointed. High, remote habitat is usually a good indicator—few deer or sheep climb so high—but in severe weather the Mountain Goat may come down into deer territory.

Bighorn Sheep

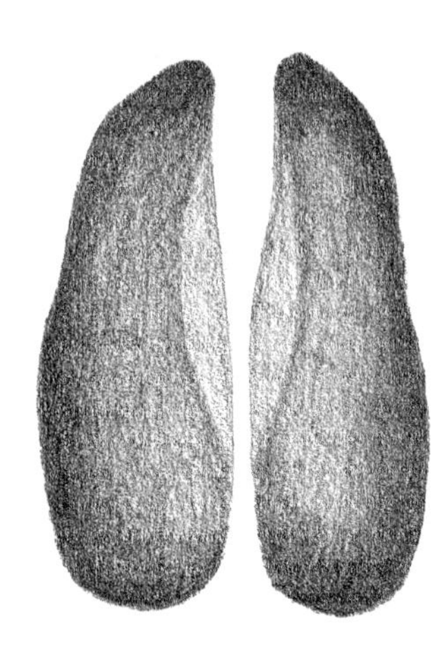

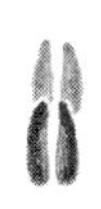

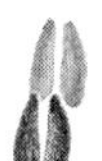

walking

Fore and Hind Prints
Length: 6.5–9 cm (2.5–3.5 in)
Width: 4.5–6.5 cm (1.8–2.5 in)

Straddle
15–30 cm (6–12 in)

Stride
Walking: 35–60 cm (14–24 in)

Size (ram>ewe)
Height: 75–110 cm (2.5–3.5 ft)
Length: 1.2–2 m (4–6.5 ft)

Weight
34–120 kg (75–270 lb)

BIGHORN SHEEP
(Mountain Sheep)

Ovis canadensis

In late fall, the loud crack of two majestic rams head-butting one another can be heard for a great distance. To watch the rut is an awe-inspiring but rare experience. The Bighorn Sheep prefers high, open meadows and scree slopes in the mountains, moving into valleys only in winter.

The squarish print is pointed toward the front. The outer edge of the hoof is hard and the inner part is soft, giving the sheep a good grip on tricky terrain. The neat alternating walking pattern is a direct or double register of the hind atop the fore. When this sheep runs, its toes splay wide. Bighorn Sheep tracks (likely several together, because this sheep likes to travel in herds) may lead you to sheep beds—hollows in the snow that are used many times and often have a large accumulation of droppings.

Similar Species: Deer (pp. 26–29) prints are more heart-shaped. The Mountain Goat (p. 32) rarely runs, enjoys craggier terrain, and its print is wider at the toe.

Horse

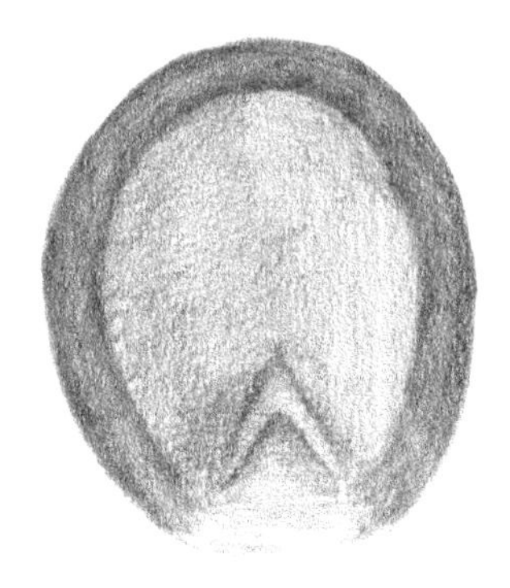

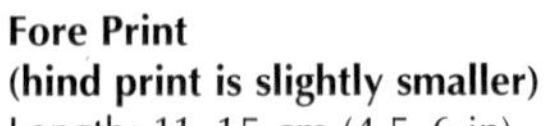

Fore Print
(hind print is slightly smaller)
Length: 11–15 cm (4.5–6 in)
Width: 11–14 cm (4.5–5.5 in)

Straddle
5–19 cm (2–7.5 in)

Stride
Walking: 43–70 cm (17–28 in)

Size
Height: to 1.8 m (6 ft)

Weight
to 680 kg (1500 lb)

walking

HORSE

Equus caballus

Wilderness adventures on horseback are a popular activity, so you can expect horse tracks to show up almost anywhere.

Unlike any other animal in this book, the Horse has just one huge toe on each foot. This toe leaves an oval print with a distinctive 'frog' (V-shaped mark) at its base. If a Horse is shod, the horseshoe shows up clearly as a firm wall at the outside of the print. Not all horses are shod, so do not expect to see this outer wall on every horse print. A typical, unhurried horse trail shows an alternating walking pattern, with the hind prints registered on or behind the slightly larger fore prints. Horses are capable of a range of speeds—up to a full gallop—but most recreational horseback riders take a more leisurely outlook on life, preferring to walk their horses and soak up the beautiful views!

Similar Species: A Caribou (p. 20) print will be smaller and show two distinct parts. On a hard surface, Bison (p. 18) prints may appear similar.

Grizzly Bear

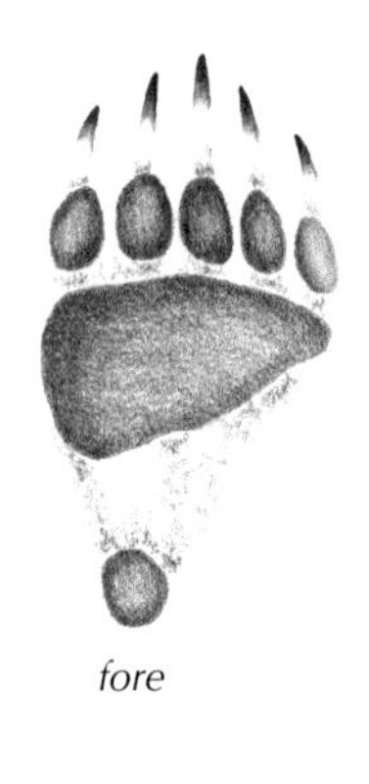

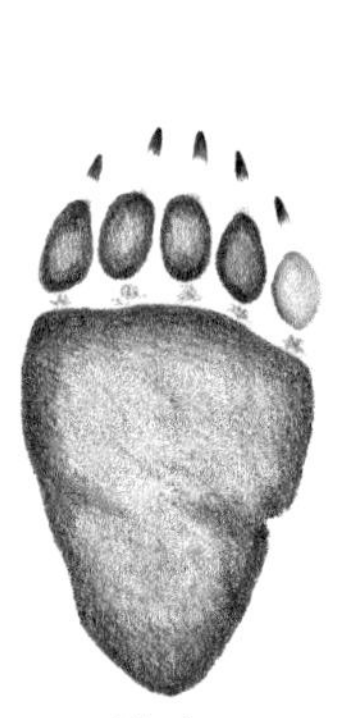

fore

hind

walking

Fore Print
Length: 13–18 cm (5–7 in)
Width: 10–15 cm (4–6 in)

Hind Print
Length: 23–30 cm (9–12 in)
Width: 13–18 cm (5–7 in)

Straddle
25–50 cm (10–20 in)

Stride
Walking: 60–100 cm (24–40 in)

Size (male>female)
Height: 90–130 cm (3–4.3 ft)
Length: 1.8–2.1 m (6–7 ft)

Weight
110–600 kg (240–1300 lb)

GRIZZLY BEAR
(Brown Bear)
Ursus arctos

The infrequently seen but magnificent Grizzly Bear symbolizes mountain wilderness for many of us. This bear prefers open country and valley bottoms, and it is sensitive to human activity. It is very sparsely distributed throughout Alberta's mountains. In winter the Grizzly Bear enters a deep slumber, so few of its tracks are seen.

Each of the Grizzly's huge prints will show four or five toes, with very long claws and a small heel pad on the fore print. A solid rear heel makes for a sturdy hind print. The toes are closely set in a line—the inner toe is the smallest. The usual walking track pattern shows the hind print registering ahead of the fore print. A slower gait results in a pattern like the Black Bear's (p. 40). The Grizzly occasionally gallops. Well-worn trails may lead to digs, trees with claw marks high up on the trunks or even the cache of a carcass. Take care if you find a cache—the unpredictable Grizzly is likely to be nearby.

Similar Species: A Black Bear's prints are smaller, with shorter claw marks and toes arranged more in an arc, and its territorial tree-scratchings are lower on the trunk.

Black Bear

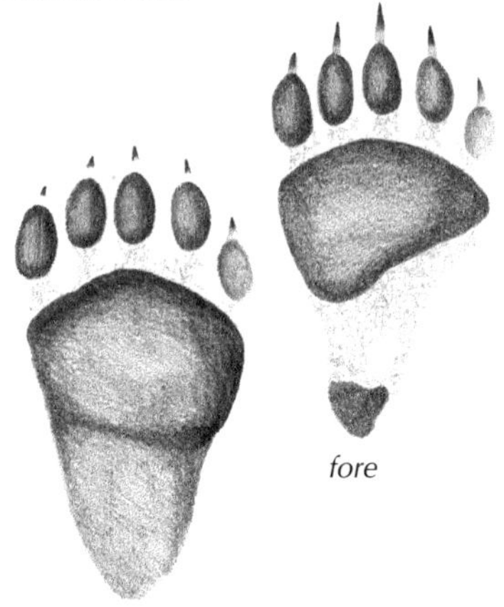

fore

hind

Fore Print
Length: 10–16 cm (4–6.3 in)
Width: 9.5–14 cm (3.8–5.5 in)

Hind Print
Length: 15–18 cm (6–7 in)
Width: 9–14 cm (3.5–5.5 in)

Straddle
23–38 cm (9–15 in)

Stride
Walking: 43–58 cm (17–23 in)

Size (male>female)
Height: 90–110 cm (3–3.5 ft)
Length: 1.5–1.8 m (5–6 ft)

Weight
90–270 kg (200–600 lb)

walking

BLACK BEAR

Ursus americanus

The Black Bear has a scattered range in forested areas throughout most of the province, but do not expect to see its tracks in winter, when it sleeps deeply. Finding fresh bear tracks can be a thrill, but take care—the bear may be just ahead. Never underestimate the potential power of a surprised bear!

Black Bear prints resemble small human prints, but they are wider and show claw marks. The small inner toe rarely registers. The forefoot's small heel pad often registers, and the hind print shows a big heel. The bear's slow walk results in a slightly pigeon-toed double register with the hind print on the fore print. More frequently, at a faster pace, the hind foot oversteps the forefoot. When a bear runs, the two hind feet register in front of the forefeet in an extended cluster. Along well-worn bear paths, look for 'digs'—patches of dug-up earth—and 'bear trees' whose scratched bark shows that this bear climbs.

Similar Species: The magnificent Grizzly Bear (p. 38), which has larger prints, is not as widespread.

Grey Wolf

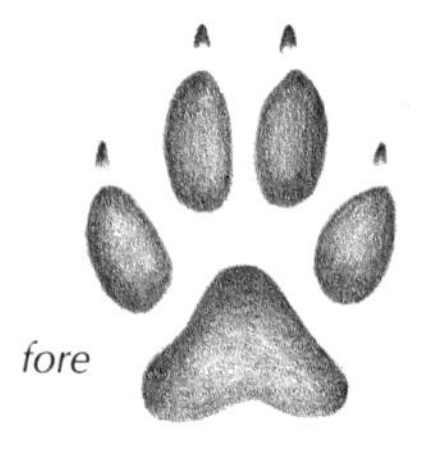

fore

hind

Fore Print
(hind print is slightly smaller)
Length: 10–14 cm (4–5.5 in)
Width: 6.5–13 cm (2.5–5 in)

Straddle
7.5–18 cm (3–7 in)

Stride
Walking: 38–80 cm (15–32 in)
Galloping: 90 cm (3 ft)
leaps to 2.7 m (9 ft)

Size (female is slightly smaller)
Height: 65–95 cm (25–37 in)
Length: 1.1–1.6 m (3.5–5.3 ft)

Weight
32–55 kg (70–120 lb)

walking

trotting

GREY WOLF
(Timber Wolf)
Canis lupus

The soulful howl of the wolf epitomizes the outdoor experience, but few people ever hear it—your best chance is in national parks or remote, undisturbed areas. The largest of the wild dogs, the Grey Wolf may travel in packs or alone, but it is rarely seen.

A wolf leaves a straight alternating track pattern of large, oval prints that each show all four claws. The smaller hind foot registers directly on the larger fore print. The lobing on the fore and hind heel pads differs. In deep snow, wolves sensibly follow their leader's trail, sometimes dragging their feet. When a wolf trots, notice how the hind print has a slight lead and falls to one side, giving an unbalanced appearance. Wolves and Coyotes (p. 46) gallop in the same way.

Similar Species: Domestic Dog (p. 44) prints, rarely as large as a wolf's, fall in a haphazard track pattern with a less direct register, and the inner toes tend to splay more. Coyote tracks are much smaller. Unclear Wolverine (p. 64) prints may appear similar.

Domestic Dog

fore

hind

walking *loping to galloping*

Fore Print (hind print is smaller)
Length: 2.5–14 cm (1–5.5 in)
Width: 2.5–13 cm (1–5 in)

Straddle
3.8–20 cm (1.5–8 in)

Stride
Walking: 7.5–80 cm (3–32 in)
Loping to Galloping: to 2.7 m (9 ft)

Size
Very variable

Weight
Very variable

DOMESTIC DOG

Canis familiaris

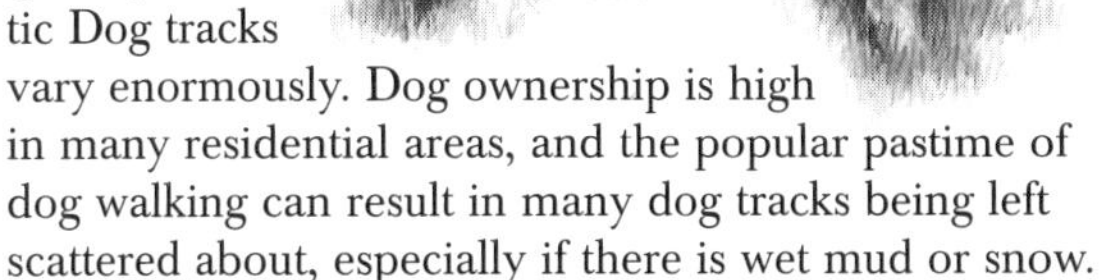

Dogs come in many shapes and sizes, from the tiny Chihuahua with its dainty feet to the robust and powerful Great Dane. Consequently, Domestic Dog tracks vary enormously. Dog ownership is high in many residential areas, and the popular pastime of dog walking can result in many dog tracks being left scattered about, especially if there is wet mud or snow.

The forefeet of the Domestic Dog, which are much larger than the hind feet and support more of the animal's weight, leave the clearest tracks. When a dog walks, the hind prints usually register ahead of or beside the fore prints. As the dog moves faster, it trots and then lopes before it gallops. In a trot or lope pattern the prints alternate fore-hind-fore-hind, whereas a gallop group shows (from back to front) fore-fore-hind-hind.

Similar Species: Keep in mind that dog tracks are usually found close to human tracks or activity. The Grey Wolf (p. 42) has large, dog-like tracks but prefers wilderness areas. The Coyote's (p. 46) more oval prints tend to splay less, and its trail is more direct. Fox (pp. 48–51) tracks may be confused with small dog tracks.

Coyote

Fore Print
(hind print is slightly smaller)
Length: 6–8 cm (2.4–3.2 in)
Width: 4–6 cm (1.6–2.4 in)

Straddle
10–18 cm (4–7 in)

Stride
Walking: 20–40 cm (8–16 in)
Trotting: 43–58 cm (17–23 in)
Galloping/Leaping:
0.8 m–3 m (2.5–10 ft)

Size (female is slightly smaller)
Height: 58–65 cm (23–26 in)
Length: 80–100 cm (32–40 in)

Weight
9–23 kg (20–50 lb)

walking or trotting

gallop group

COYOTE
(Brush Wolf, Prairie Wolf)
Canis latrans

This widespread, adaptable canine prefers open grasslands or woodlands. On its own, with a mate or in a family pack, it hunts rodents and larger prey. A Coyote occasionally develops an interesting cooperative relationship with a Badger (p. 76); you might find their tracks together where they have been digging for ground squirrels (p. 96). If you find a Coyote den—usually a wide-mouthed tunnel leading into a nesting chamber—do not bother the family, or the female will move her pups.

The hind print is slightly smaller than the oval fore print, and the less-triangular hind heel pad rarely registers clearly. Usually just the middle two toes register claw marks. The Coyote typically walks or trots in an alternating pattern; the walk has a wider straddle, and the trotting trail is often very straight. When it gallops, the Coyote's hind feet fall in front of its forefeet; the faster it goes, the straighter the gallop group. The Coyote's tail, which hangs down, leaves a dragline in deep snow.

Similar Species: A Domestic Dog's (p. 44) less-oval prints splay more, and its trail is erratic. Foot hairs blur Red Fox (p. 48) prints (usually smaller).

Red Fox

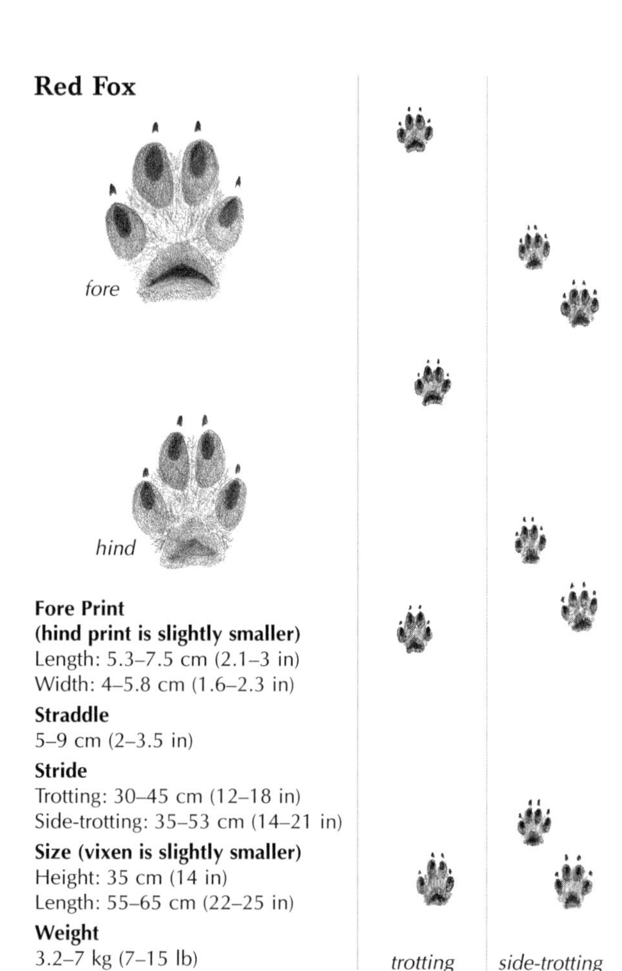

Fore Print
(hind print is slightly smaller)
Length: 5.3–7.5 cm (2.1–3 in)
Width: 4–5.8 cm (1.6–2.3 in)

Straddle
5–9 cm (2–3.5 in)

Stride
Trotting: 30–45 cm (12–18 in)
Side-trotting: 35–53 cm (14–21 in)

Size (vixen is slightly smaller)
Height: 35 cm (14 in)
Length: 55–65 cm (22–25 in)

Weight
3.2–7 kg (7–15 lb)

RED FOX

Vulpes vulpes

Very adaptable and intelligent, this beautiful and notoriously cunning fox is found throughout Alberta in a variety of habitats, from forests to open areas.

Abundant foot hair allows only parts of the toes and heel pads to register, with no fine detail. The horizontal or slightly curved bar across the fore heel pad is diagnostic. A trotting Red Fox leaves a distinctive straight alternating trail—the hind print direct registers on the wider fore print. When the fox side-trots, its print pairs show the hind print to one side of the fore print in typical canid fashion. This fox gallops like the Coyote (p. 46). The faster the gallop, the straighter the gallop group.

Similar Species: Other canid prints lack the bar across the fore heel pad. Domestic Dog (p. 44) prints can be of similar size, but with a shorter stride and a less direct trail. Swift Fox (p. 50) tracks are smaller. Small Coyote prints are similar but have a wider straddle, and the toe marks are more bulbous.

Kit Fox

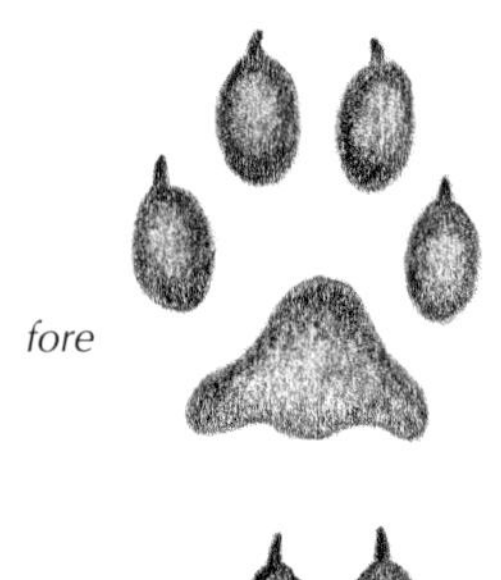

fore

hind

trotting

Fore and Hind Prints
Length: 2.8–4.5 cm (1.1–1.8 in)
Width: 2.8–3.8 cm (1.1–1.5 in)

Straddle
5–10 cm (2–4 in)

Stride
Walking/Trotting: 18–25 cm (7–10 in)

Size
Height: 30 cm (1 ft)
Length with tail: 60–80 cm (24–32 in)

Weight
1.4–2.7 kg (3–6 lb)

SWIFT FOX

Vulpes velox

This small, shy fox of the plains and arid regions of southern Alberta tends to hide away. Secretive and solitary, its nocturnal activity makes it a rare sight, but tracks can reveal its presence.

The Swift Fox's preference for dusty areas means that its tracks will seldom be very clear, because dirt will often fall back into the prints. If you find unclear prints, pay attention to general track characteristics, such as whether the prints are in a typical dog-family trotting pattern. The track's dimensions may help identify which species made them.

Similar Species: The Red Fox (p. 48) has different heel pads (with a bar across them); in general, its prints are larger and less clear (because of thick fur), its stride is longer, and its straddle is narrower. Domestic Dog (p. 44) prints can be of similar size, but with a shorter stride and a less direct trail. Domestic Cat (p. 58) and Bobcat (p. 56) prints lack claw marks and have larger, less symmetrical heel pads.

Mountain Lion

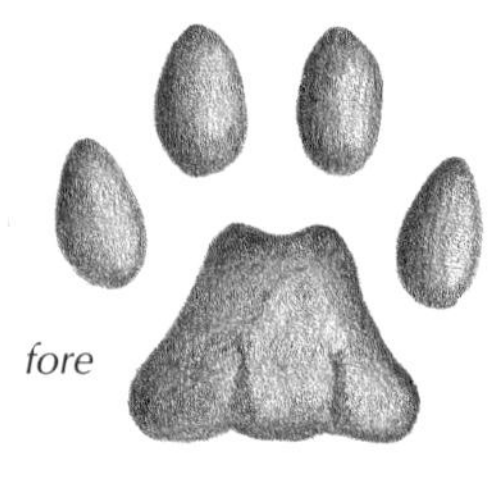

fore

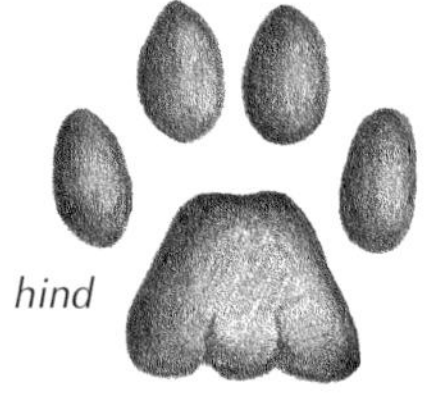

hind

Fore Print
(hind print is slightly smaller)
Length: 7.5–11 cm (3–4.5 in)
Width: 8.5–12 cm (3.3–4.8 in)

Straddle
20–30 cm (8–12 in)

Stride
Walking: 33–80 cm (13–32 in)
Bounding: to 3.7 m (12 ft)

Size
Height: 65–80 cm (25–32 in)
Length: 3.5–5 ft (1.1–1.5 m)

Weight
32–90 kg (70–200 lb)

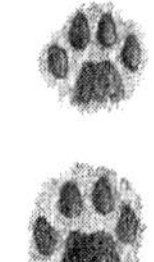

walking (fast)

MOUNTAIN LION (Cougar, Puma, Panther)

Puma concolor

Shy, elusive and nocturnal, the Mountain Lion is spread widely but sparsely because of its need for a big home territory. Finding its tracks is usually the best that you can hope for. No longer common in much of its historic range, this cat and its tracks may be seen in mountain areas and parts of southern Alberta.

Mountain Lion prints tend to be wider than long. The retractable claws never register. Thick foot fur enlarges the print in winter and may stop the two lobes on the front of the heel pad from registering clearly. In the walking gait, the hind print direct registers or double registers on the larger fore print. As the pace increases, the hind print tends to fall ahead of the fore print. In snow, the thick, long tail may leave a dragline that can blur print detail. A Mountain Lion seldom gallops, but it is capable of long bounds. Also look for partly buried scat and kills covered for later eating.

Similar Species: Bobcat (p. 56) prints may be confused with juvenile Mountain Lion prints. The Lynx (p. 54) has prints with a narrower straddle and a shorter stride, and it does not drag its tail or sink as deep in snow. Grey Wolf (p. 42) and Coyote (p. 46) tracks show claw marks.

Lynx

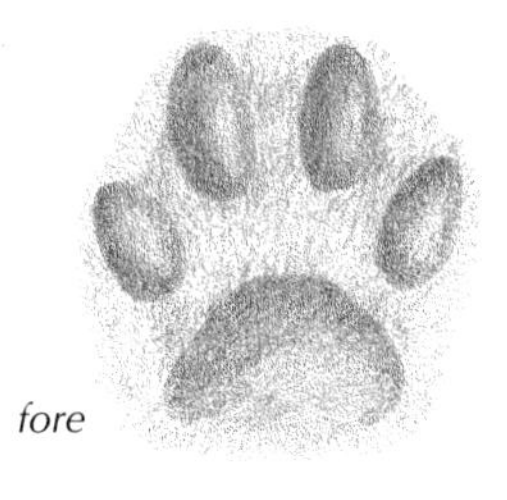
fore

hind

Fore Print
(hind print is slightly smaller)
Length: 9–11 cm (3.5–4.5 in)
Width: 9–12 cm (3.5–4.8 in)

Straddle
15–23 cm (6–9 in)

Stride
Walking: 30–70 cm (12–28 in)

Size
Length: 75–90 cm (2.5–3 ft)

Weight
7–14 kg (15–30 lb)

walking

LYNX
Lynx canadensis

This large cat is a thrill to see, but it is sensitive to human interference. The elusive Lynx is abundant only in remote, undisturbed, dense forests. With huge feet and a relatively lightweight body, it stays on top of the snow as it pursues its main prey, the Snowshoe Hare (p. 80).

This cautious walker leaves a neat alternating track pattern, the hind print direct registered on top of the fore print. Thick fur on the feet often results in prints that are big, round depressions with no detail. In deeper snow, 'handles' off to the rear may extend the print. However deep the snow, this cat sinks no more than 20 cm (8 in), and it rarely drags its feet. The Lynx is more likely to bound than to run. Curiosity results in a meandering trail that may lead to a partially buried food cache.

Similar Species: The Mountain Lion (p. 52), with similar prints—but clearer and with a wider straddle—sinks deeper in snow. Bobcat (p. 56) prints are smaller; its trail may show draglines. Fisher (p. 66) prints may not show the fifth toe—look for mustelid habits. Canid (pp. 42–51) prints show claw marks and their length exceeds their width.

Bobcat

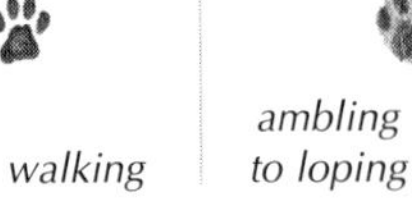

Fore Print
(hind print is slightly smaller)
Length: 4.5–6.5 cm (1.8–2.5 in)
Width: 4.5–6.5 cm (1.8–2.5 in)

Straddle
10–18 cm (4–7 in)

Stride
Walking: 20–40 cm (8–16 in)
Running: 1.2–2.4 m (4–8 ft)

Size (female is slightly smaller)
Height: 50–55 cm (20–22 in)
Length: 65–75 cm (25–30 in)

Weight
7–16 kg (15–35 lb)

BOBCAT (Wildcat)

Lynx rufus

The Bobcat, a stealthy and usually nocturnal hunter found across southern Alberta, is seldom seen. Very adaptable, it can leave tracks anywhere from wild mountainsides to chaparral and even residential areas.

A walking Bobcat's hind feet usually register directly on its larger fore prints. As the Bobcat picks up speed, its trail becomes an ambling pattern of paired prints, the hind leading the fore. At even greater speeds it leaves four-print groups in a lope pattern. The fore prints often exhibit asymmetry. The front part of the heel pad has two lobes and the rear part has three. In deep snow the Bobcat's feet leave draglines. Half-buried scat along the Bobcat's meandering trail marks its territory.

Similar Species: Lynx (p. 54) or juvenile Mountain Lion (p. 52) prints can be similar. Large Domestic Cats (p. 58) have similar prints but a shorter stride and a narrower straddle. Canid (pp. 42–51) prints are narrower than long and will show claw marks, and the front of the footpad will have just one lobe.

Domestic Cat

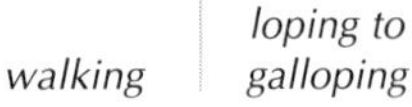

Fore Print
(hind print is slightly smaller)
Length: 2.5–4 cm (1–1.6 in)
Width: 2.5–4.5 cm (1–1.8 in)

Straddle
6–11 cm (2.4–4.5 in)

Stride
Walking: 13–20 cm (5–8 in)
Loping/Galloping:
35–80 cm (14–32 in)

Size (male>female)
Height: 50–55 cm (20–22 in)
Length with tail: 75 cm (30 in)

Weight
3–6 kg (6.5–13 lb)

DOMESTIC CAT
(House Cat)
Felis catus

The tracks of the familiar and abundant Domestic Cat can show up almost any place where there are people. Abandoned cats may roam farther afield, and these 'feral cats' lead a pretty wild and independent existence. Domestic Cats come in many shapes, sizes and colours.

As with all felines, a Domestic Cat's fore print and slightly smaller hind print both show four toe pads. Its retractable claws, kept clean and sharp for catching prey, do not register. Cat prints usually show a slight asymmetry, with one toe leading the others. The Domestic Cat makes a neat alternating walking track pattern, usually in direct register, as one would expect from this animal's fastidious nature. When a cat picks up speed, it leaves four-print clusters, the hind feet registering in front of the forefeet.

Similar Species: A small Bobcat (p. 56) may leave tracks similar to a very large Domestic Cat's. Fox (pp. 48–51) and Domestic Dog (p. 44) prints show claw marks.

Raccoon

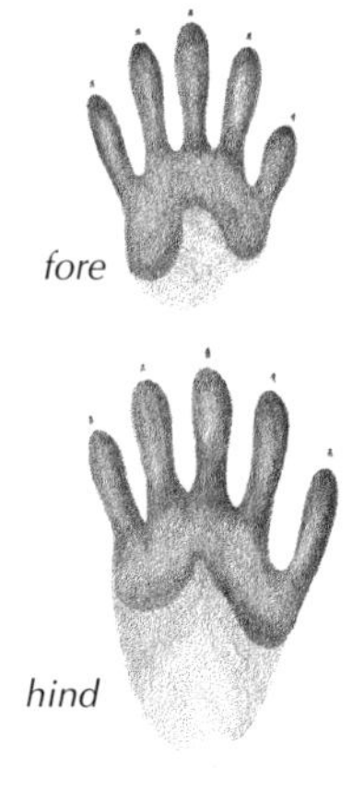

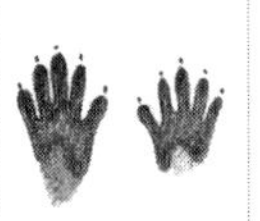

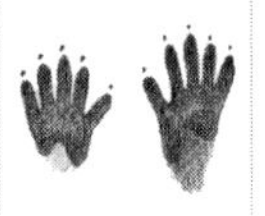

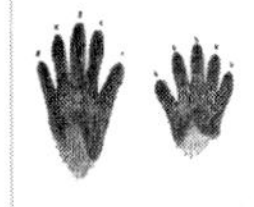

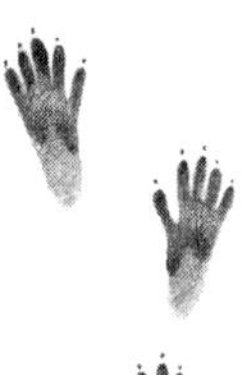

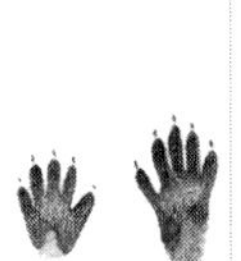

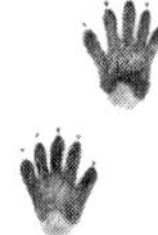

walking

bounding group

Fore Print
Length: 5–7.5 cm (2–3 in)
Width: 4.5–6.5 cm (1.8–2.5 in)

Hind Print
Length: 6–9.5 cm (2.4–3.8 in)
Width: 5–6.5 cm (2–2.5 in)

Straddle
8.5–15 cm (3.3–6 in)

Stride
Walking: 20–45 cm (8–18 in)
Bounding: 38–65 cm (15–25 in)

Size (female is slightly smaller)
Length: 60–95 cm (24–37 in)

Weight
5–16 kg (11–35 lb)

RACCOON

Procyon lotor

Found in southern Alberta, the inquisitive Raccoon is adored by some people for its distinctive face mask, yet disliked for its boundless curiosity—often demonstrated with residential garbage cans. A good place to look for its tracks is near water at low elevations. The Raccoon likes to rest in trees. It usually dens up for the colder months.

The Raccoon's unusual print, showing five well-formed toes, looks like a human handprint; its small claws make dots. Its highly dexterous forefeet rarely leave heel prints, but its hind prints, which are generally much clearer, do show heels. The Raccoon's peculiar walking track pattern shows the left fore print next to the right hind print (or just in front) and vice versa. On the rare occasions when a Raccoon is out in deep snow, it may use a direct-registering walk. The Raccoon occasionally bounds, leaving clusters with the hind prints in front of the fore prints.

Similar Species: Raccoon prints are normally distinctive, but, in deep snow, Fisher (p. 66), River Otter (p. 62) or Woodchuck (p. 94) tracks may look similar.

River Otter

fore

hind

Fore Print
Length: 6.5–9 cm (2.5–3.5 in)
Width: 5–7.5 cm (2–3 in)

Hind Print
Length: 7.5–10 cm (3–4 in)
Width: 5.8–8.5 cm (2.3–3.3 in)

Straddle
10–23 cm (4–9 in)

Stride
Loping: 30–70 cm (12–27 in)

Size
(female is two-thirds the size of male)
Length with tail: 90–130 cm (3–4.3 ft)

Weight
4.5–11 kg (10–25 lb)

loping (fast)

RIVER OTTER

Lontra canadensis

No animal knows how to have more fun than a River Otter. If you are lucky enough to watch one at play, you will not soon forget the experience. Widespread and well-adapted for the aquatic environment, this otter lives near water; an otter in the forest is usually on its way to another waterbody. Expect to see a wealth of evidence of an otter's presence along the waterbodies in its home territory. The River Otter loves to slide in snow, often down riverbanks, leaving troughs nearly 30 cm (1 ft) wide. Lacking snow, it rolls and slides on grass and mud.

In soft mud the River Otter's five-toed feet, especially the hind ones, register evidence of webbing. The inner toe is set slightly apart from the rest. If the forefoot's metacarpal pad registers, it lengthens the print. Very variable, otter trails usually show a typical mustelid 2×2 loping. However, with faster gaits they can sometimes show three- and four-print groups. The thick, heavy tail often leaves a dragline.

Similar Species: Other mustelid trails do not show conspicuous tail drag. The Marten (p. 68) has similar-sized prints. Mink (p. 70) prints are about half the size. The Fisher (p. 66) usually has hairy feet, with forefeet larger than hind feet.

Wolverine

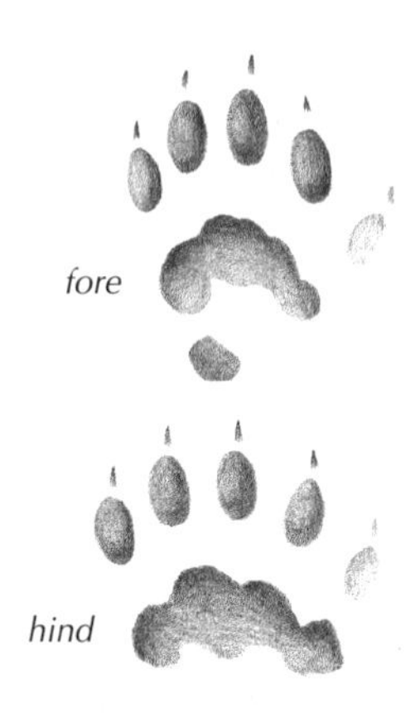

Fore Print
Length with heel: 10–19 cm (4–7.5 in)
Width: 10–13 cm (4–5 in)

Hind Print
Length: 9–10 cm (3.5–4 in)
Width: 10–13 cm (4–5 in)

Straddle
18–23 cm (7–9 in)

Stride
Walking: 7.5–30 cm (3–12 in)
Running: 25–100 cm (10–40 in)

Size (female is slightly smaller)
Height: 40 cm (16 in)
Length: 80–120 cm (2.7–3.8 ft)

Weight
8–21 kg (18–45 lb)

WOLVERINE
Gulo gulo

The reputation of the robust and powerful Wolverine, one of the largest members of the weasel family, has earned it many nicknames, such as 'Skunk Bear' and 'Indian Devil.' Each Wolverine's need for a large territory (which can encompass diverse, generally wooded habitats) and pristine wilderness has resulted in a scattered distribution throughout much of the animal's range.

As with some other mustelids, the Wolverine's inner toe (of five) rarely shows in the print. Though the forefoot registers a small heel pad, the hind foot rarely does. The Wolverine's low, squat shape results in a host of erratic typical mustelid patterns: an alternating walking pattern, the typical 2×2 loping pattern with its print pairs, and the common loping pattern of three- and four-print groups.

Similar Species: Most other mustelids (pp. 62, 66–77) make much smaller prints with less-erratic track patterns. When only four toes register, Wolverine tracks may be mistaken for Grey Wolf (p. 42) or Domestic Dog (p. 44) tracks, but the pad shapes are quite different.

Fisher

walking

Fore Print
Length: 5.3–10 cm (2.1–4 in)
Width: 5.3–8.5 cm (2.1–3.3 in)

Hind Print
Length: 5.3–7.5 cm (2.1–3 in)
Width: 5–7.5 cm (2–3 in)

Straddle
7.5–18 cm (3–7 in)

Stride
Walking: 18–35 cm (7–14 in)
2×2 loping: 30–130 cm (1–4.3 ft)
Loping: 30–90 cm (1–3 ft)

Size (male>female)
Length with tail:
85–100 cm (34–40 in)

Weight
1.4–5.5 kg (3–12 lb)

FISHER (Black Cat)

Martes pennanti

This agile hunter is comfortable both on the ground and in the trees of mixed hardwood forests. The Fisher's speed and eager hunting antics make for exciting tracking as it races up trees and along the ground in its quest for squirrels. It is one of the few predators capable of killing and eating Porcupines (p. 88).

All five toes may register, but the small inner toe frequently does not. Only the forefoot has a small heel pad that can show up in the print. The Fisher occasionally walks, making a direct-registering alternating track pattern, but it more often 2×2 lopes in typical mustelid fashion, leaving angled print pairs with the hind print direct registered on the fore print. Loping, its most common gait, produces three- and four- print groups (see the River Otter, p. 62). The patterns often vary within a short distance. Not associated with water, Fishers have been misnamed!

Similar Species: Male Marten (p. 68) tracks may be confused with a small female Fisher's, but Martens weigh less and leave shallower prints. Otters have larger hind feet than forefeet. When only four toes register, Fisher prints may look like a Bobcat's (p. 56).

Marten

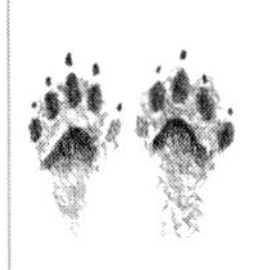

Fore and Hind Prints
Length: 4.5–6.5 cm (1.8–2.5 in)
Width: 3.8–7 cm (1.5–2.8 in)

Straddle
6.5–10 cm (2.5–4 in)

Stride
Walking: 10–23 cm (4–9 in)
2×2 loping: 23–120 cm (9–46 in)

Size (male>female)
Length with tail:
53–70 cm (21–28 in)

Weight
0.7–1.3 kg (1.5–2.8 lb)

walking

MARTEN
(American Sable)

Martes americana

This aggressive predator inhabits northern Alberta's coniferous and mixed-wood forests.

The Marten seldom leaves a clear print—often just four toes register, and the heel pad is undeveloped. The hairiness of the feet in winter often blurs pad detail, especially from the poorly developed palm pads. In the Marten's alternating walking pattern, the hind foot registers on the fore print. In 2×2 loping, the hind prints fall on the fore prints to form slightly angled print pairs in a typical mustelid pattern. The Marten's loping track patterns may appear as three- or four-print clusters (see the River Otter, p. 62). Follow the criss-crossing trails—if a Marten has scrambled up a tree, look for a sitzmark where it has jumped down.

Similar Species: Size and habitat are often key to distinguishing Marten, Fisher (p. 66) and Mink (p. 70) tracks. Female Fisher prints may resemble a large male Marten's but will be clearer. Male Mink prints overlap in size with small female Marten prints, but the Mink rarely climbs trees and (unlike the Marten) is usually found near water.

Mink

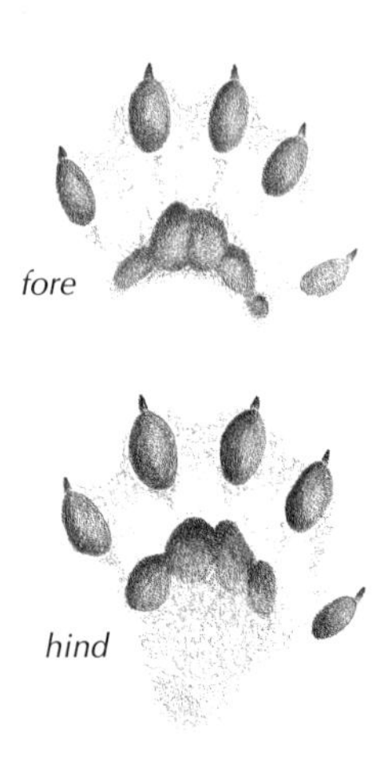

Fore and Hind Prints
Length: 3.3–5 cm (1.3–2 in)
Width: 3.3–4.5 cm (1.3–1.8 in)

Straddle
5.3–9 cm (2.1–3.5 in)

Stride
Walking/2×2 loping: 20–90 cm (8–36 in)

Size (male>female)
Length with tail: 48–70 cm (19–28 in)

Weight
0.7–1.6 kg (1.5–3.5 lb)

MINK

Mustela vison

The lustrous Mink, which is widespread throughout the province, prefers watery habitats surrounded by brush or forest. At home as much on land as in water, this nocturnal hunter can be exciting to track. Like the River Otter (p. 62), the Mink sometimes slides in snow, carving out a trough up to 15 cm (6 in) wide.

The Mink's fore print shows five (perhaps four) toes, with five loosely connected palm pads in an arc, but the hind print shows only four palm pads. The metacarpal pad of the forefoot rarely registers, but the furred heel of the hind foot may register, lengthening the hind print. The Mink prefers the typical mustelid 2×2 loping gait, making consistently spaced, slightly angled double prints. Its diverse track patterns also include alternating walking, loping with three- and four-print groups (like the River Otter) and bounding (like a rabbit or hare, pp. 80–85).

Similar Species: Small Martens (p. 68) may have similar prints, but without a consistent 2×2 loping gait; they do not live near water. Weasel (pp. 72–75) tracks are similar but smaller. Bobcat (p. 56) prints may resemble four-toed Mink prints, but without claw marks or lobed palm pads, and they will not be in a bounding pattern.

Weasels

all weasels

Long-tailed Weasel

Fore and Hind Prints
Length: 2.8–4.5 cm (1.1–1.8 in)
Width: 2–2.5 cm (0.8–1 in)

Straddle
4.5–7 cm (1.8–2.8 in)

Stride
2×2 loping: 24–110 cm (9.5–43 in)

Size (male>female)
Length with tail: 30–55 cm (12–22 in)

Weight
85–340 g (3–12 oz)

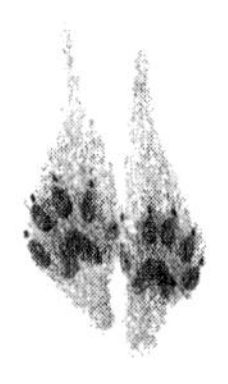

Long-tailed

LONG-TAILED WEASEL

Mustela frenata

Weasels are active, year-round hunters with an avid appetite for rodents. The Long-tailed Weasel is the largest of the three weasels in the province.

Following a weasel's tracks can reveal much about the activities of the nimble creature. Tracks are most evident in winter, when weasels frequently burrow into the snow or pursue rodents into their holes. Some weasel trails may lead you up a tree. Weasels sometimes take to water. To identify the weasel species, pay close attention to the straddle, stride and loping patterns, and note the distribution and habitat. The usual weasel gait is a 2×2 lope, leaving a trail of paired prints. A weasel's light weight and small, hairy feet result in pad detail that is often unclear, especially in snow. Even with clear tracks, the inner (fifth) toe rarely registers.

The Long-tailed Weasel's typical 2×2 lope shows an irregular stride—sometimes short and sometimes long—with no consistent behaviour. Like the Mink (p. 70), this weasel may bound like a rabbit or hare (p. 80–85).

Similar Species: A large male Short-tailed Weasel's (p. 74) tracks may be the same size as a small female Long-tailed Weasel's. The Least Weasel (p. 74), with smaller tracks, occupies some of the same habitats.

Weasels

Short-tailed Weasel

Fore and Hind Prints
Length: 2–3.3 cm (0.8–1.3 in)
Width: 1.3–1.5 cm (0.5–0.6 in)

Straddle
2.5–5.3 cm (1–2.1 in)

Stride
2×2 loping: 23–90 cm (9–36 in)

Size (male>female)
Length with tail:
20–35 cm: (8–14 in)

Weight
30–170 g (1–6 oz)

Least Weasel

Fore and Hind Prints
Length: 1.3–2 cm (0.5–0.8 in)
Width: 1–1.3 cm (0.4–0.5 in)

Straddle
2–3.8 cm (0.8–1.5 in)

Stride
2x2 loping: 13–50 cm (5–20 in)

Size (male>female)
Length with tail:
17–23 cm (6.5–9 in)

Weight
37–65 g (1.3–2.3 oz)

Short-tailed 2×2 loping

Least

SHORT-TAILED WEASEL
(Ermine, Stoat)

Mustela erminea

The Short-tailed Weasel is smaller than the Long-tailed Weasel (p. 72) but larger than the Least Weasel (below). It prefers woodlands and meadows up to higher elevations but does not favour wetlands or dense coniferous forests.

This weasel's 2×2 loping tracks may fall in clusters, with alternating short and long strides.

Similar Species: A small female Long-tailed's tracks may be the same size as a large male Short-tailed's.

LEAST WEASEL

Mustela nivalis

The Least Weasel, found throughout the province, is the smallest weasel, with the least-clear tracks.

Its tracks may be found around wetlands and in open woodlands and fields.

Similar Species: A small female Short-tailed Weasel's (above) tracks may resemble a large male Least Weasel's, but the Short-tailed's habitats differ.

Badger

fore

hind

walking

Fore Print
(hind print is slightly shorter)
Length: 6.5–7.5 cm (2.5–3 in)
Width: 5.8–7 cm (2.3–2.8 in)

Straddle
10–18 cm (4–7 in)

Stride
Walking: 15–30 cm (6–12 in)

Size
Length: 53–90 cm (21–36 in)

Weight
6–11 kg (13–25 lb)

BADGER

Taxidea taxus

The squat shape and unmistakable face of this bold animal are most often seen in open grasslands, but the Badger also ventures into higher country. It is found throughout much of Alberta. Thick shoulders and forelegs, coupled with long claws, make it a powerful digger. Look for the Badger's tracks in spring and autumn snow—unlike most other mustelids, it likes to den up in a hole for the coldest months of winter.

When a Badger walks, the alternating track pattern shows a double register, with the hind print sometimes falling just behind (or sometimes slightly in front of) the fore print. All five toes on each foot register. A Badger's long claws are evident in the pigeon-toed tracks that it leaves as it waddles along; the forefoot claws are longer than the hind-foot ones. In deep snow, the ploughing action of the Badger's wide, low body often wipes out track detail.

Similar Species: A Porcupine (p. 88) trail in snow may be similar but will show draglines made by the tail and quills, and it will likely lead up a tree, not to a hole.

Striped Skunk

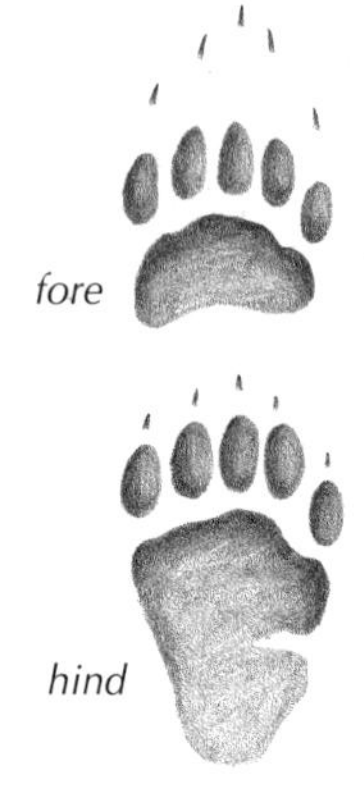

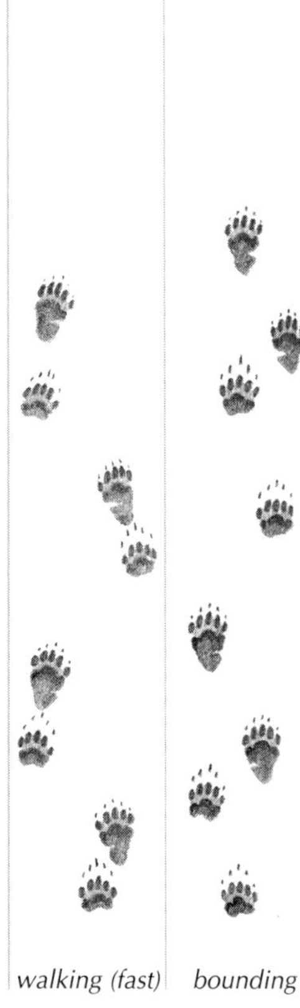

Fore Print
Length: 3.8–5.6 cm (1.5–2.2 in)
Width: 2.5–3.8 cm (1–1.5 in)

Hind Print
Length: 3.8–6.5 cm (1.5–2.5 in)
Width: 2.5–3.8 cm (1–1.5 in)

Straddle
7–11 cm (2.8–4.5 in)

Stride
Walking/Bounding:
6.5–20 cm (2.5–8 in)

Size
Length with tail:
50–80 cm (20–32 in)

Weight
2.7–6.5 kg (6–14 lb)

STRIPED SKUNK

Mephitis mephitis

This striking skunk is notorious for its vile smell, and the lingering odour is often the best sign of its presence. Widespread throughout the province in a diversity of habitats, it prefers lower elevations. The Striped Skunk dens up in winter, coming out on warmer days and in spring.

Forefeet and hind feet each have five toes. The long claws on the forefeet often register. The smooth palm pads and small heel pads leave surprisingly small prints. A skunk mostly walks—with such a potent smell for its defence, and those memorable black and white stripes, it rarely needs to run. Note that its trail rarely shows any consistent pattern, but an alternating walking pattern may be evident. The greater a skunk's speed, the more the hind foot oversteps the fore. If it runs, its trail consists of clumsy, closely set four-print groups. In snow it drags its feet.

Similar Species: There are no other wild skunk species in the province. Similar-sized mustelid (pp. 66–77) tracks will be farther apart than those resulting from a skunk's shuffling gait, and skunk prints do not overlap.

Snowshoe Hare

fore

hind

Fore Print
Length: 5–7.5 cm (2–3 in)
Width: 3.8–5 cm (1.5–2 in)

Hind Print
Length: 10–15 cm (4–6 in)
Width: 5–9 cm (2–3.5 in)

Straddle
15–20 cm (6–8 in)

Stride
Hopping: 25–130 cm (0.8–4.3 ft)

Size
Length: 30–53 cm (12–21 in)

Weight
0.9–1.8 kg (2–4 lb)

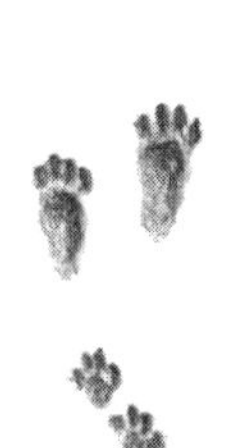

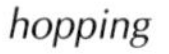

hopping

SNOWSHOE HARE
(Varying Hare)
Lepus americanus

This hare is well known for its colour change from summer brown to winter white and for its huge hind feet, which enable it to 'float' on top of snow. Widespread throughout the province, it frequents brushy areas in forests, which provide good cover from the Lynx (p. 54) and the Coyote (p. 46), its most likely predators. Hares are most active at night.

The Snowshoe Hare's most common track pattern is a hopping one, with triangular four-print groups; they can be quite long if the hare moves quickly. In winter, heavy fur on the hind feet (much larger than the forefeet) thickens the toes, which can splay out to further distribute the hare's weight on snow. Hares make well-worn runways that are often used as escape runs. You may encounter a resting hare, because hares do not live in burrows. Twigs and stems neatly cut at a 45° angle also indicate this hare's presence.

Similar Species: The Mountain Cottontail (p. 84) has much smaller prints. The White-tailed Jackrabbit (p. 82) of the south splays its hind toes less.

White-tailed Jackrabbit

fore

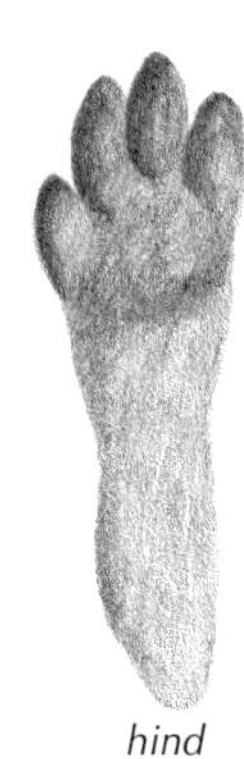

hind

hopping

Fore Print
Length: 6.5–10 cm (2.5–4 in)
Width: 3.3–4.3 cm (1.3–1.7 in)

Hind Print
Length: 9–17 cm (3.5–6.5 in)
Width: 3.8–6.5 cm (1.5–2.5 in)

Straddle
11–18 cm (4.5–7 in)

Stride
Hopping: 30–300 cm (1–10 ft)

Size
Length: 58–65 cm (23–25 in)

Weight
2.3–4 kg (5–9 lb)

WHITE-TAILED JACKRABBIT

Lepus townsendii

This hare frequents open country in southern Alberta. An athletic animal, the White-tailed Jackrabbit can reach the impressive speed of 70 km/h (45 mph). Because of its nocturnal and solitary habits and its wariness of predators, it is infrequently seen.

Both fore and hind prints show four toes; the hind foot may often register a long heel. When it hops, this hare creates print groups in a triangular pattern; as it speeds up, these print groups spread out considerably. Following a hare's trail could lead you to its 'form'—a depression where it rests—or an urgent zigzag pattern that indicates where the hare fled from danger. With its strong hind legs, it is capable of leaping up to 6 m (20 ft) to avoid pursuers.

Similar Species: The Snowshoe Hare (p. 80) splays its hind toes more, takes shorter leaps and requires dense cover. Coyote (p. 46) prints resemble heel-less jackrabbit prints, but the pattern is very different. The Mountain Cottontail (p. 84) has a much smaller print cluster and a shorter stride.

Mountain Cottontail

fore

hind

Fore Print
Length: 2.5–3.8 cm (1–1.5 in)
Width: 2–3.3 cm (0.8–1.3 in)

Hind Print
Length: 7.5–9 cm (3–3.5 in)
Width: 2.5–3.8 cm (1–1.5 in)

Straddle
10–13 cm (4–5 in)

Stride
Hopping: 18–90 cm (0.6–3 ft)

Size
Length: 30–43 cm (12–17 in)

Weight
0.6–1.4 kg (1.3–3 lb)

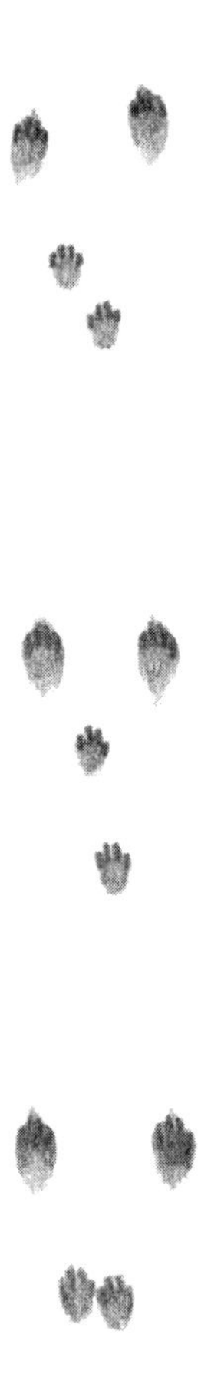

hopping

MOUNTAIN COTTONTAIL (Nuttall's Cottontail)

Sylvilagus nuttallii

This abundant rabbit is found only in the south of the province. Preferring brushy areas in grasslands and cultivated areas, it might be encountered in dense vegetation, hiding from predators such as the Bobcat (p. 56) and the Coyote (p. 46). Largely nocturnal, the Mountain Cottontail might be seen at dawn or dusk and on darker days.

As with other rabbits and hares, this rabbit's most common track pattern is a triangular grouping of four prints, with the larger hind prints (which can appear pointed) falling in front of the fore prints (which may overlap). The hairiness of the toes will hide any pad detail. If you follow this rabbit's trail, you may be startled as it flies out from its 'form,' a depression in the ground in which it rests.

Similar Species: The Snowshoe Hare (p. 80) has larger prints, especially the hind ones. The White-tailed Jackrabbit (p. 82) leaves much larger print clusters and has longer strides.

Pika

fore

hind

Fore Print
Length: 2 cm (0.8 in)
Width: 1.5 cm (0.6 in)

Hind Print
Length: 2.5–3 cm (1–1.2 in)
Width: 1.5–2 cm (0.6–0.8 in)

Straddle
6.5–9 cm (2.5–3.5 in)

Stride
Walking/Bounding: 10–25 cm (4–10 in)

Size
Length: 17–22 cm (6.5–8.5 in)

Weight
110–170 g (4–6 oz)

bounding

PIKA
(Cony, Rock Rabbit)
Ochotona princeps

When hiking high up in the mountains, you are more likely to hear the squeak of this cousin of the rabbits than to see it, because the Pika is quick to disappear under the rocks when alarmed. Confined to areas of high elevation, the Pika rarely leaves good tracks, because it prefers exposed rocky areas on mountain slopes. Its tracks are most likely to be found in spring, on patches of snow or in mud. A more conspicuous sign of the Pika's presence is its little hay piles, set to dry in the sun for the winter ahead. During this time, the Pika feeds on its stored food as it remains active under the snow.

The fore print usually shows five toes, although the fifth toe does not always register, but the hind print shows only four. The prints may appear in an erratic alternating pattern or in three- and four-print bounding groups.

Similar Species: The Pika's high-mountain habitat means that its prints are rarely confused with rabbit or hare (pp. 80–85) prints, though track patterns may be similar.

Porcupine

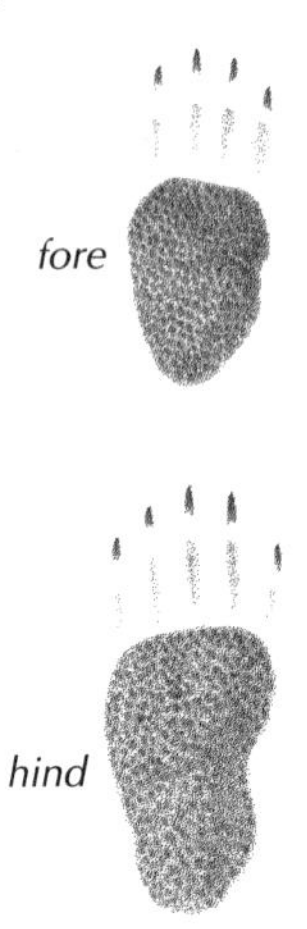

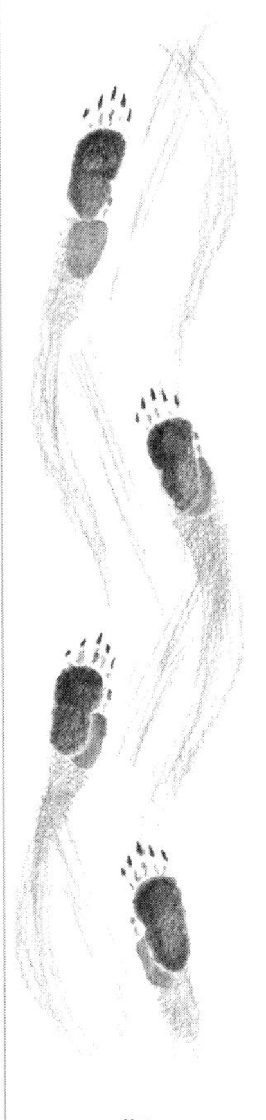
walking

Fore Print
Length: 5.8–8.5 cm (2.3–3.3 in)
Width: 3.3–4.8 cm (1.3–1.9 in)

Hind Print
Length: 7–10 cm (2.8–4 in)
Width: 3.8–5 cm (1.5–2 in)

Straddle
14–23 cm (5.5–9 in)

Stride
Walking: 13–25 cm (5–10 in)

Size
Length with tail: 65–100 cm (25–40 in)

Weight
4.5–13 kg (10–28 lb)

PORCUPINE

Erethizon dorsatum

This notorious rodent rarely runs—its many long quills are a formidable defence. Widespread across Alberta, the Porcupine prefers forests, but it can also be seen in more-open areas.

The Porcupine's preferred pigeon-toed, waddling gait leaves an alternating track pattern, with the hind print registered on or slightly in front of the shorter fore print. Look for long claw marks on all prints. The fore print shows four toes, and the hind print shows five. Clear prints may show the unusual pebbly surface of the solid heel pads, but a Porcupine's tracks are often scratch-marked by its heavy, spiny tail. In deeper snow this squat animal drags its feet, and it may leave a trough with its body. A Porcupine's trail might lead you to a tree, where this animal spends much of its time feeding; if so, look for chewed bark or nipped twigs on the ground.

Similar Species: The Badger (p. 76) makes pigeon-toed prints, but its tracks don't show tail drag, and it doesn't climb trees.

Beaver

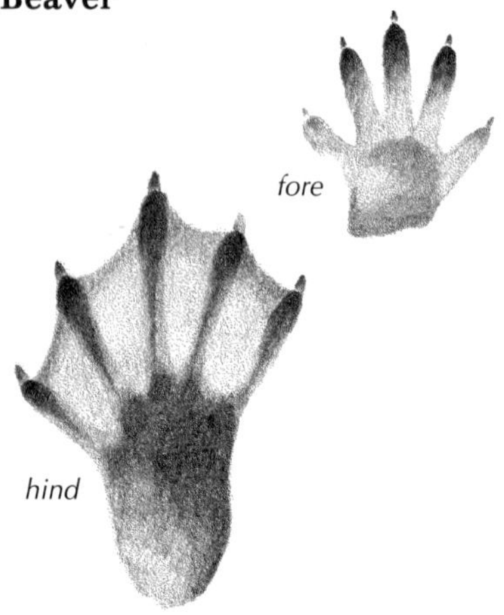

Fore Print
Length: 6.5–10 cm (2.5–4 in)
Width: 5–9 cm (2–3.5 in)

Hind Print
Length: 13–18 cm (5–7 in)
Width: 8.5–13 cm (3.3–5.3 in)

Straddle
15–28 cm (6–11 in)

Stride
Walking: 7.5–17 cm (3–6.5 in)

Size
Length with tail: 90–120 cm (3–4 ft)

Weight
13–34 kg (28–75 lb)

BEAVER

Castor canadensis

Few animals leave as many signs of their presence as the Beaver, North America's largest rodent and a common sight around water. Look for the conspicuous dams and lodges—capable of changing the local landscape—and the stumps of felled trees. Inspect trunks gnawed clean of bark for marks of the Beaver's huge incisors. Scent mounds marked with castoreum, a strong-smelling yellowish fluid that Beavers produce, also indicate recent activity.

Check the large hind prints for signs of webbing and broad toenails. The nail of the second inner toe usually does not register, and it is rare for all five toes on each foot to do so. Irregular foot placement in the alternating walking gait may produce a direct register or a double register. The Beaver's thick, scaly tail may mar its tracks, as can the branches that it drags about for construction and food. Repeated path use results in well-worn trails.

Similar Species: The Beaver's many signs, including its large hind prints, minimize confusion. Muskrat (p. 92) prints are smaller.

Muskrat

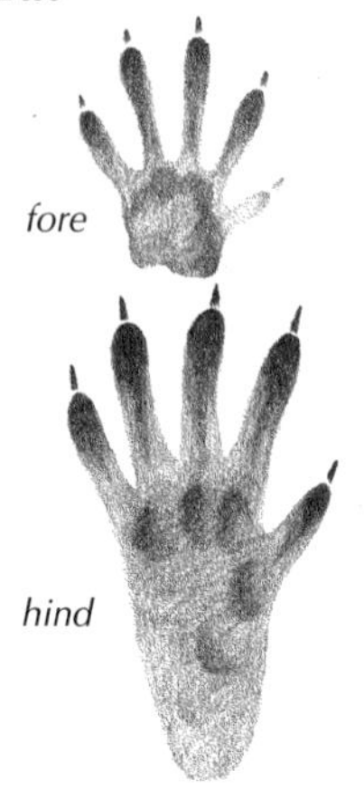

Fore Print
Length: 2.8–3.8 cm (1.1–1.5 in)
Width: 2.8–3.8 cm (1.1–1.5 in)

Hind Print
Length: 4–8 cm (1.6–3.2 in)
Width: 3.8–5.3 cm (1.5–2.1 in)

Straddle
7.5–13 cm (3–5 in)

Stride
Walking: 7.5–13 cm (3–5 in)
Running: to 30 cm (1 ft)

Size
Length with tail: 40–65 cm (16–25 in)

Weight
0.9–1.8 kg (2–4 lb)

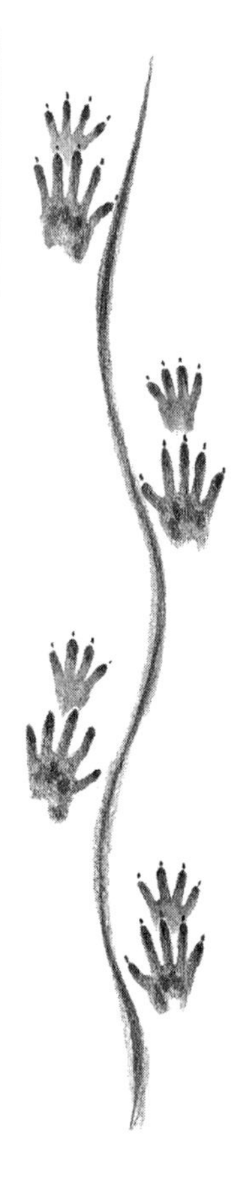

walking

MUSKRAT

Ondatra zibethicus

Like the Beaver (p. 90), this rodent can be found throughout the province, wherever there is water. Beavers are very tolerant of Muskrats and even allow them to live in parts of their lodges. Active year-round, the Muskrat leaves plenty of signs. It digs extensive networks of burrows, often undermining riverbanks, so do not be surprised if you suddenly fall into a hidden hole! Also look for small lodges in the water and beds of vegetation on which the Muskrat rests, suns and feeds in summer.

The small fifth (inner) toe of the forefoot rarely registers. Stiff hairs that aid in swimming may create a 'shelf' around the five well-formed toes of the hind print. The common alternating walking pattern shows print pairs that alternate from side to side; the hind print is just behind the fore print or slightly overlaps it. In snow, a Muskrat's feet drag, and its tail leaves a sweeping dragline in its tracks.

Similar Species: Few animals share this water-loving rodent's habits. The Beaver makes larger tracks and leaves many other signs.

Woodchuck

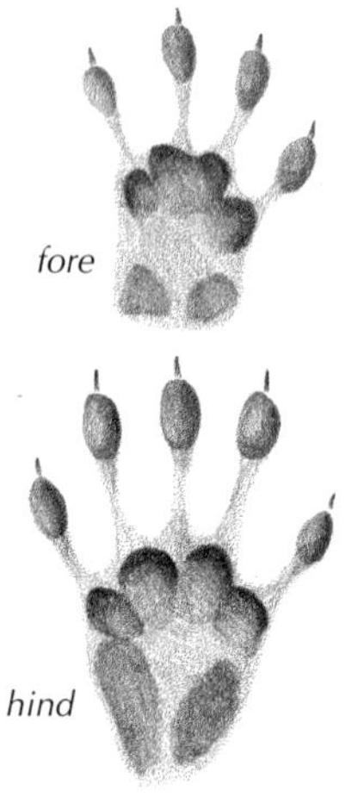

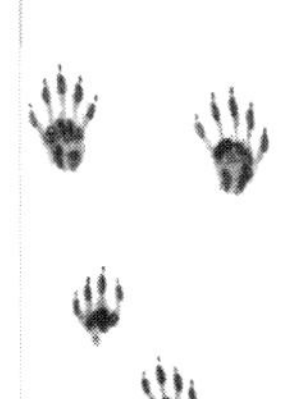

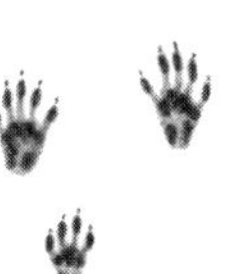

walking *bounding*

Fore and Hind Prints
Length: 4.5–7 cm (1.8–2.8 in)
Width: 2.5–5 cm (1–2 in)

Straddle
8.5–15 cm (3.3–6 in)

Stride
Walking: 5–15 cm (2–6 in)
Bounding: 15–35 cm (6–14 in)

Size (male>female)
Length with tail:
50–65 cm (20–25 in)

Weight
2.5–5.5 kg (5.5–12 lb)

WOODCHUCK (Whistle Pig, Groundhog, Marmot)

Marmota monax

This robust member of the squirrel family is a common sight in open woodlands and adjacent areas throughout northern Alberta. Always on the watch for predators, but not too troubled by humans, the Woodchuck never wanders far from its burrow. This marmot hibernates during winter but emerges in early spring; look for tracks in late spring snowfalls and in mud around burrow entrances.

A Woodchuck's fore print shows four toes, three palm pads and two heel pads (not always evident). The hind print shows five toes, four palm pads and two poorly registering heel pads. The Woodchuck usually leaves an alternating walking pattern, with the hind print registered on the fore print. When a Woodchuck runs from danger, it makes groups of four prints, hind ahead of fore.

Similar Species: The Yellow-bellied Marmot (*Marmota flaviventris*) of southernmost Alberta has similar prints. A small Raccoon's (p. 60) bounding track pattern will show five-toed fore prints. Large Muskrat (p. 92) tracks will show a tail dragline and will always be near water.

Thirteen-lined Ground Squirrel

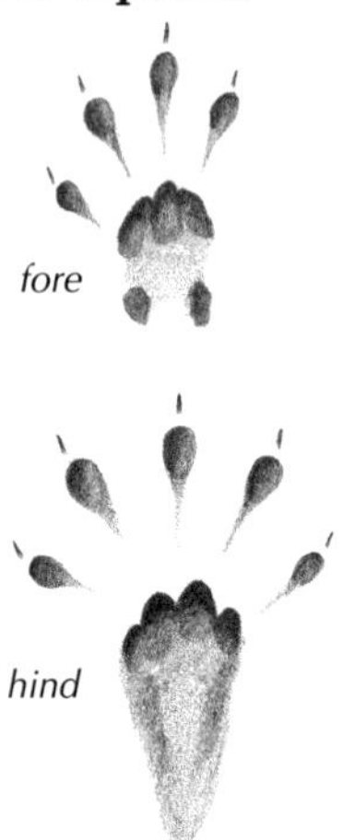

Fore Print
Length: 2.5–3.3 cm (1–1.3 in)
Width: 1.3–2.5 cm (0.5–1 in)

Hind Print
Length: 2.8–3.8 cm (1.1–1.5 in)
Width: 2–3.3 cm (0.8–1.3 in)

Straddle
5.8–9 cm (2.3–3.5 in)

Stride
Bounding: 18–50 cm (7–20 in)

Size
Length with tail: 18–30 cm (7–12 in)

Weight
110–280 g (4–10 oz)

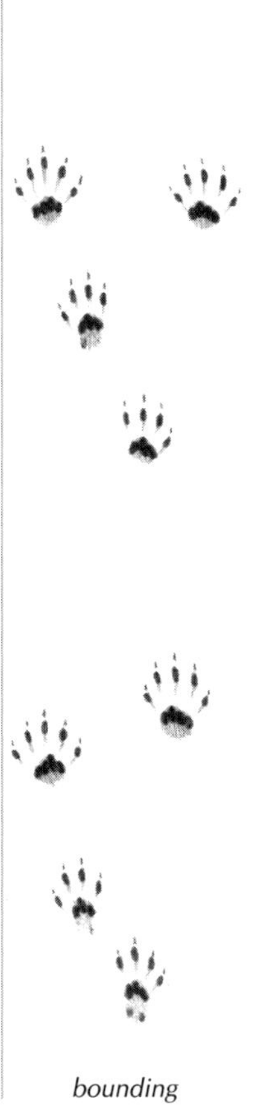

bounding

THIRTEEN-LINED GROUND SQUIRREL (Striped Gopher)

Spermophilus tridecemlineatus

With its many stripes and dots, this adorable ground squirrel is an attractive sight. It inhabits shortgrass areas, primarily in the warmer southern parts of the province.

This animal's tracks may be evident near its many burrow entrances, in mud or in late or early snowfalls—it hibernates in winter. The small fifth toe of the forefoot rarely shows in the print, but the two heel pads sometimes show. The larger hind prints show five toes. Both forefeet and hind feet have long claws that frequently register. Usually seen scurrying around, ground squirrels leave a typical squirrel track pattern, with the hind prints registering ahead of the fore prints, which are usually placed diagonally.

Similar Species: Franklin's Ground Squirrel (*S. franklinii*) is larger. Chipmunk (p. 98) tracks are smaller. Tree squirrel (pp. 100–103) tracks have a more square-shaped bounding group.

Least Chipmunk

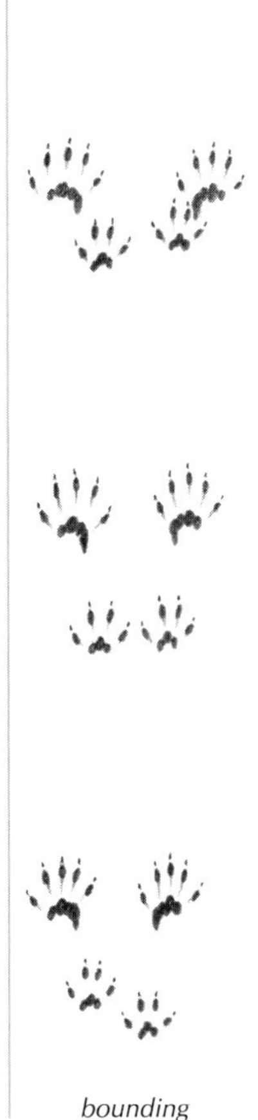

Fore Print
Length: 2–2.5 cm (0.8–1 in)
Width: 1–2 cm (0.4–0.8 in)

Hind Print
Length: 1.8–3.3 cm (0.7–1.3 in)
Width: 1.3–2.3 cm (0.5–0.9 in)

Straddle
5–8 cm (2–3.2 in)

Stride
Bounding: 18–38 cm (7–15 in)

Size
Length with tail: 18–23 cm (7–9 in)

Weight
28–70 g (1–2.5 oz)

LEAST CHIPMUNK

Tamias minimus

This delightful chipmunk is found in a variety of habitats, from dry sagebrush flats to forests, and it is bold enough to be a popular visitor in campgrounds. You are more likely to see or hear this rodent, which is highly active during summer months, than to notice its tracks. Chipmunks sleep deeply in winter, but they occasionally venture out on milder days.

Chipmunks are so light that their tracks rarely show fine details. The forefeet each have four toes, and the hind feet have five. Chipmunks run on their toes, so the two heel pads of the forefeet seldom register; the hind feet have no heel pads. Their erratic track patterns, like those of many of their cousins, show the hind feet registered in front of the forefeet. A chipmunk trail often leads to extensive burrows. Piles of nutshells on rocks are a further indication of a chipmunk's recent presence.

Similar Species: In the foothills and mountains of Alberta, the Yellow-pine Chipmunk (*T. amoenus*) has similar tracks. Mid-winter tracks of this kind are more likely those of a tree squirrel (pp. 100–103), whose tracks will usually be larger. Mouse (pp. 112–115) tracks are smaller.

Red Squirrel

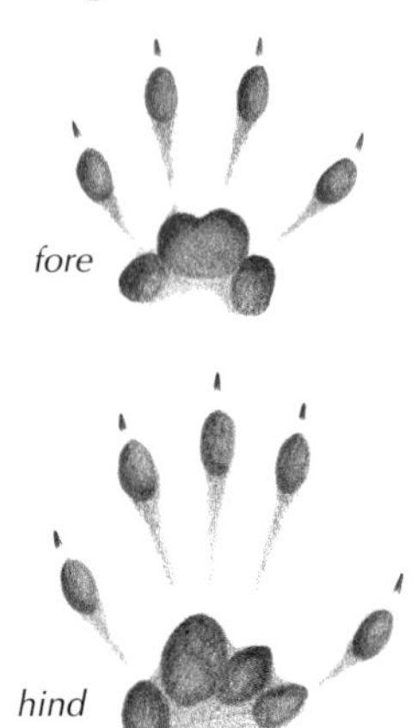

Fore Print
Length: 2–3.8 cm (0.8–1.5 in)
Width: 1.3–2.5 cm (0.5–1 in)

Hind Print
Length: 3.8–5.8 cm (1.5–2.3 in)
Width: 2–3.3 cm (0.8–1.3 in)

Straddle
7.5–11 cm (3–4.5 in)

Stride
Bounding: 20–75 cm (8–30 in)

Size
Length with tail:
23–38 cm (9–15 in)

Weight
57–260 g (2–9 oz)

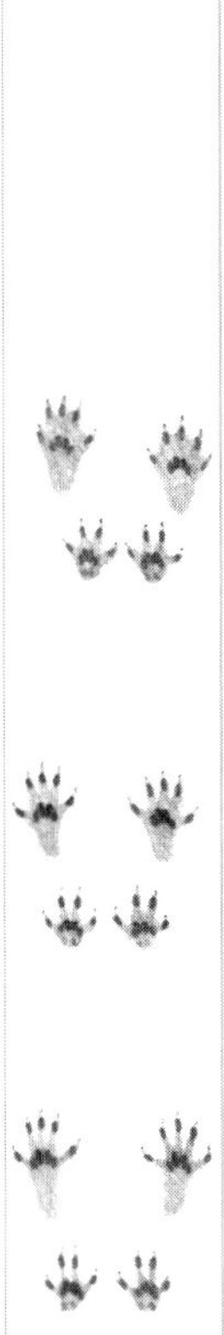

bounding

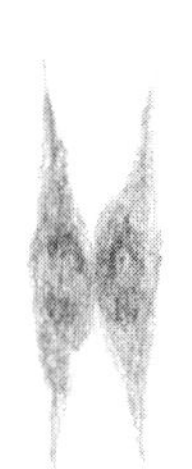

bounding (deep snow)

RED SQUIRREL (Pine Squirrel, Chickaree)

Tamiasciurus hudsonicus

When you enter a Red Squirrel's territory, the inhabitant greets you with a loud, chattering call. Another obvious sign of this forest dweller, which is found in forested areas of the province, is its large middens—piles of cone scales and cores left beneath trees—that indicate favourite feeding sites.

Active all year in its small territory, a Red Squirrel will leave an abundance of trails that lead from tree to tree or down a burrow. This energetic animal mostly bounds, leaving four-print groups, the hind prints falling in front of the fore prints (which tend to be side by side, but not always). Four toes show on each fore print, and five on each hind print. The heels often do not register when a squirrel moves quickly. In deeper snow the prints merge to form pairs of diamond-shaped tracks.

Similar Species: Chipmunk (p. 98) and Northern Flying Squirrel (p. 102) tracks are in a similar pattern, but smaller and with a narrower straddle.

Northern Flying Squirrel

fore

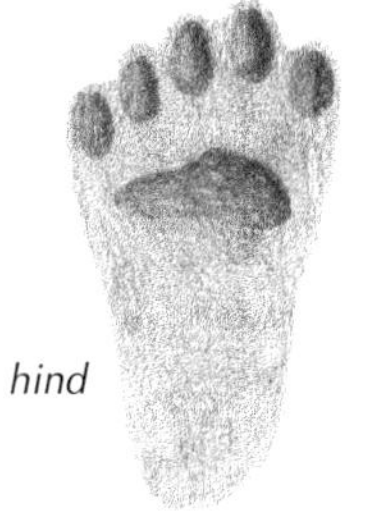
hind

Fore Print
Length: 1.3–2 cm (0.5–0.8 in)
Width: 1.3 cm (0.5 in)

Hind Print
Length: 3.3–4.5 cm (1.3–1.8 in)
Width: 2 cm (0.8 in)

Straddle
7.5–9.5 cm (3–3.8 in)

Stride
Bounding: 28–75 cm (11–30 in)

Size
Length with tail: 23–30 cm (9–12 in)

Weight
110–180 g (4–6.5 oz)

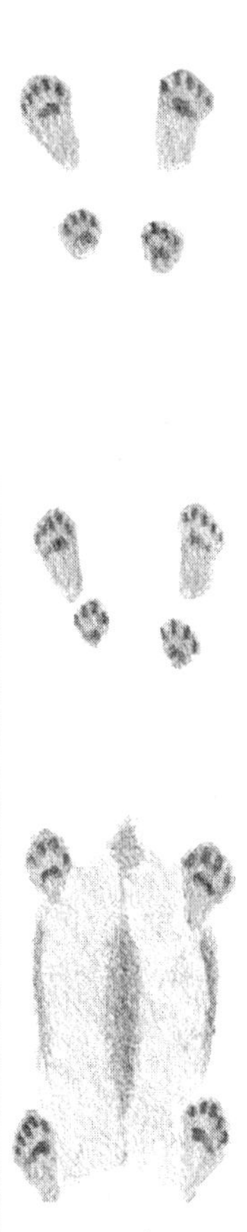
sitzmark into bounding

NORTHERN FLYING SQUIRREL

Glaucomys sabrinus

This soft-furred brown acrobat lives in coniferous and mixed forests throughout the province, excluding only the southern arid regions. It prefers widely spaced forests, where it can glide from tree to tree by night, using the membranous flaps of skin between its forelegs and hind legs. Up to 10 Northern Flying Squirrels will den up together in a tree cavity for warmth in winter.

Because of its gliding, this squirrel does not leave as many tracks as most other squirrels do. Evidence is scarce in summer, but in winter you may come across a sitzmark (the distinctive pattern made when it landed in the snow) and a short bounding trail, made as it rushed off to the nearest tree or to do some quick foraging. The bounding track pattern is typical of squirrels and other rodents, but with the hind feet registering only slightly in front of the forefeet—though often all four feet register in a row.

Similar Species: The Red Squirrel (p. 100), usually with larger prints, rarely leaves a sitzmark, but unclear tracks in deep snow can be indistinguishable. Chipmunk (p. 98) prints are smaller, with a narrower straddle.

Bushy-tailed Woodrat

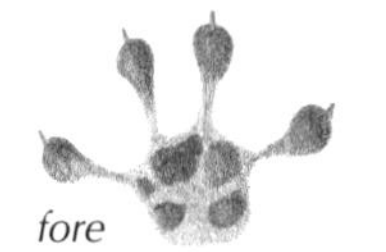

Fore Print
Length: 1.5–2 cm (0.6–0.8 in)
Width: 1–1.3 cm (0.4–0.5 in)

Hind Print
Length: 2.5–3.8 cm (1–1.5 in)
Width: 1.5–2 cm (0.6–0.8 in)

Straddle
5.8–7 cm (2.3–2.8 in)

Stride
Walking: 4.5–7.5 cm (1.8–3 in)
Bounding: 13–20 cm (5–8 in)

Size
Length with tail:
28–48 cm (11–19 in)

Weight
200–600 g (7–21 oz)

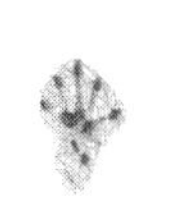
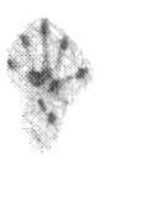

walking

bounding

BUSHY-TAILED WOODRAT (Packrat)

Neotoma cinerea

The Bushy-tailed Woodrat is found along the southern and southwestern edges of Alberta, especially in rugged terrain and mountain forests. Tracking one of these nocturnal rodents can be very rewarding: The trail might lead you to its distinctive, massed nest, which can be 1.5 m (5 ft) across and may be in an abandoned building. This animal is a curious hoarder that brings home all manner of objects, thereby serving as a selective wilderness garbage collector.

Four toes show on the fore print and five on the hind. The short claws rarely register. A woodrat often walks in an alternating fashion, with the hind print direct registering on the fore print. This woodrat frequently bounds as well, leaving a pattern of four prints, with the larger hind print in front of the diagonally placed fore prints. The stride tends to be short relative to the size of the prints.

Similar Species: The Norway Rat (p. 106) has similar prints but is usually found close to human activity. Indistinct Red Squirrel (p. 100) prints appear similar but usually larger. Woodchuck (p. 94) prints are similar but much larger.

Norway Rat

fore

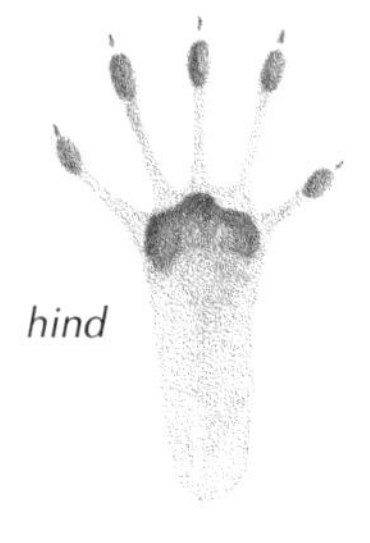

hind

Fore Print
Length: 1.8–2 cm (0.7–0.8 in)
Width: 1.3–1.8 cm (0.5–0.7 in)

Hind Print
Length: 2.5–3.3 cm (1–1.3 in)
Width: 2–2.5 cm (0.8–1 in)

Straddle
5–7.5 cm (2–3 in)

Stride
Walking: 3.8–9 cm (1.5–3.5 in)
Bounding: 23–50 cm (9–20 in)

Size
Length with tail: 33–48 cm (13–19 in)

Weight
200–510 g (7–18 oz)

walking

NORWAY RAT
(Brown Rat)

Rattus norvegicus

Despite the efforts at rat control, this despised rat is frequently found in parts of southern Alberta, almost anywhere that humans have decided to build their homes. Not entirely dependent on people, it may live in the wild as well.

The fore print shows four toes, and the hind print shows five. When it bounds, this colonial rat leaves four-print groups, hind prints in front of the diagonally placed fore prints. Sometimes one of the hind feet direct registers on a fore print, creating a three-print group. This rat more commonly leaves an alternating walking pattern with the larger hind prints close to or overlapping the fore prints; the hind heel does not show. The tail often leaves a dragline in snow. Rats live in groups, so you may find many trails together, often leading to their 2-cm (5-in) wide burrows.

Similar Species: The Bushy-tailed Woodrat (p. 104) leaves similar prints, but it is seldom associated with human activity, except in abandoned buildings. Mouse (pp. 112–115) prints are much smaller, with different track patterns. Chipmunk (p. 98) tracks are similar but smaller, and the track patterns differ. Red Squirrel (p. 100) tracks show distinctive squirrel traits.

Northern Pocket Gopher

fore

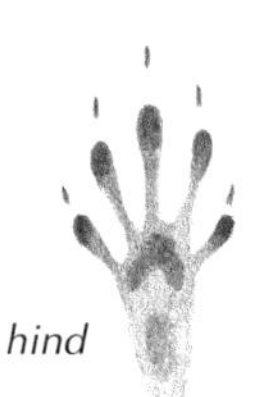

hind

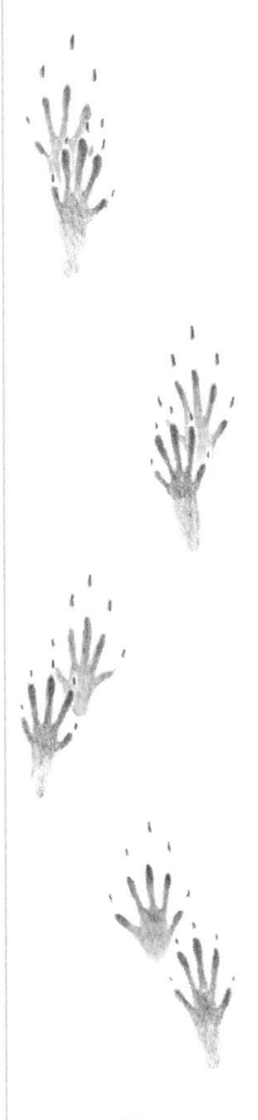

walking

Fore Print
Length: 2.5 cm (1 in)
Width: 1.5 cm (0.6 in)

Hind Print
Length: 2–2.5 cm (0.8–1 in)
Width: 1.3 cm (0.5 in)

Straddle
3.8–5 cm (1.5–2 in)

Stride
Walking: 3.3–5 cm (1.3–2 in)

Size (male>female)
Length: with tail: 15–23 cm (6–9 in)

Weight
80–140 g (2.8–5 oz)

NORTHERN POCKET GOPHER

Thomomys talpoides

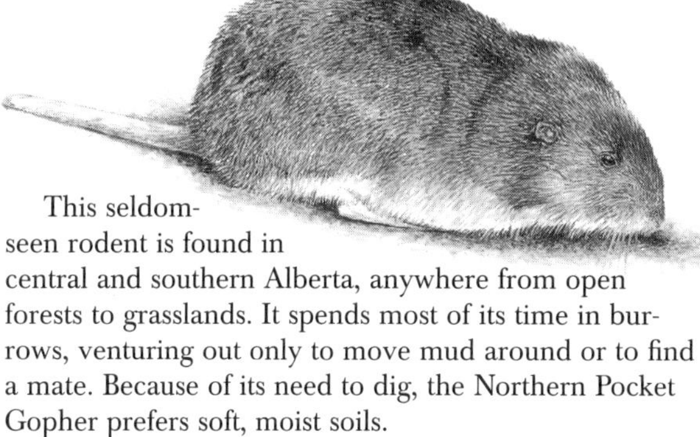

This seldom-seen rodent is found in central and southern Alberta, anywhere from open forests to grasslands. It spends most of its time in burrows, venturing out only to move mud around or to find a mate. Because of its need to dig, the Northern Pocket Gopher prefers soft, moist soils.

By far the best signs of Northern Pocket Gopher activity are the muddy mounds and tunnel cores—they are especially evident just after spring thaw. Each mound marks the entrance to a burrow, which is always blocked up with a plug. Search around the mounds to find tracks. Each foot has five toes. Though the forefeet have long, well-developed claws for digging, the prints rarely show this much detail. Pocket gophers usually walk, leaving an alternating track pattern in which the hind prints fall on or slightly behind the fore prints.

Similar Species: The Northern Pocket Gopher's tracks are associated with its distinctive burrows, leaving little room for confusion.

Meadow Vole

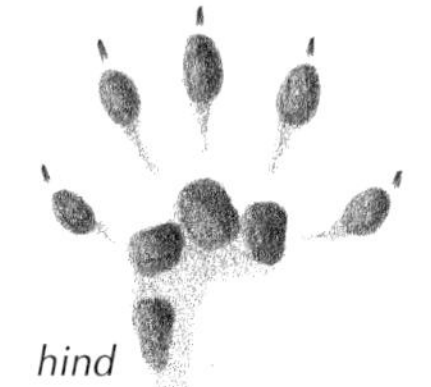

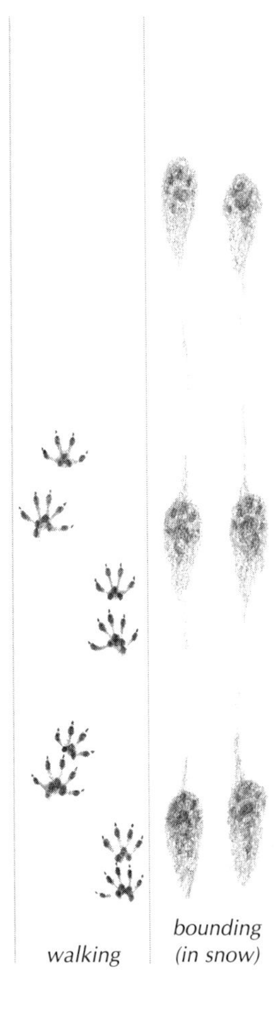

Fore Print
Length: 0.8–1.3 cm (0.3–0.5 in)
Width: 0.8–1.3 cm (0.3–0.5 in)

Hind Print
Length: 0.8–1.5 cm (0.3–0.6 in)
Width: 0.8–1.5 cm (0.3–0.6 in)

Straddle
3.3–5 cm (1.3–2 in)

Stride
Walking/Trotting:
3.3–7.5 cm (1.3–3 in)
Bounding: 10–20 cm (4–8 in)

Size
Length with tail:
14–20 cm (5.5–8 in)

Weight
14–70 g (0.5–2.5 oz)

MEADOW VOLE (Field Mouse)

Microtus pennsylvanicus

With so many vole species in Alberta, positive track identification is next to impossible, but note that the Meadow Vole prefers damp or wet habitats.

When clear (which is seldom), vole fore prints show four toes, and hind prints show five. A vole's walk and trot both leave a paired alternating track pattern with a hind print occasionally direct registered on a fore print. Voles usually opt for a faster bounding in which the hind prints register on the fore prints. They lope quickly across open areas, creating a three-print pattern. In winter, voles stay under the snow; after the thaw, look for distinctive piles of cut grass from their ground nests. The bark at the bases of shrubs may show tiny teeth marks left by gnawing. In summer, well-used vole paths appear as little runways in the grass.

Similar Species: Other common voles have similar tracks. The Southern Red-backed Vole (*Cletherionomys gapperi*) likes boggy areas and damp forests. The Heather Vole (*Phenacomys intermedius*) inhabits northern coniferous forests. The Long-tailed Vole (*M. longicaudis*) likes the drier parts of the west. The Deer Mouse (p. 112) leaves four-print bounding groups.

Deer Mouse

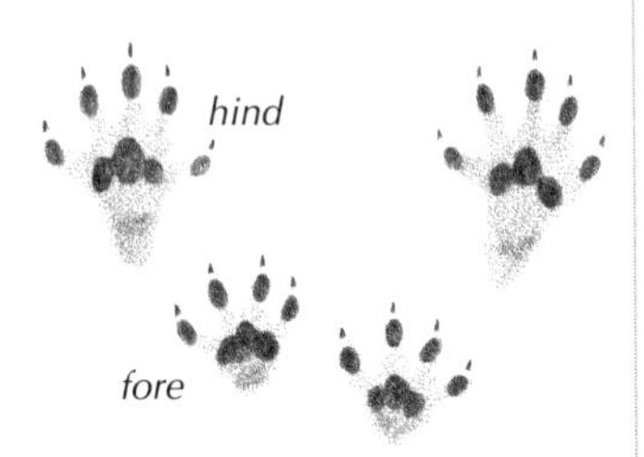

bounding group

Fore Print
Length: 0.8–1 cm (0.3–0.4 in)
Width: 0.8–1 cm (0.3–0.4 in)

Hind Print
Length: 0.8–1.3 cm (0.3–0.5 in)
Width: 0.8–1 cm (0.3–0.4 in)

Straddle
3.6–4.5 cm (1.4–1.8 in)

Stride
Bounding: 13–30 cm (5–12 in)

Size
Length with tail:
15–30 cm (6–12 in)

Weight
14–35 g (0.5–1.3 oz)

bounding

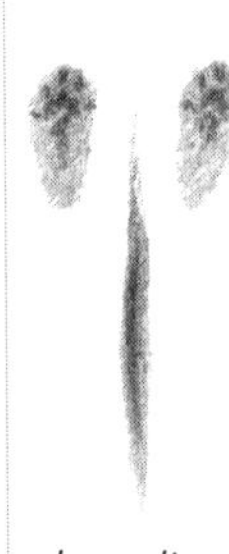

bounding (in snow)

DEER MOUSE

Peromyscus maniculatus

The highly adaptable Deer Mouse—one of the province's most abundant mammals—lives anywhere from arid valleys all the way up to alpine meadows. It is seldom seen, because it is nocturnal. The Deer Mouse may enter buildings in winter, where it will stay active.

In perfect, soft mud, the fore prints each show four toes, three palm pads and two heel pads, and the hind prints show five toes and three palm pads; the heel pads rarely register. Bounding tracks, most noticeable in snow, show the hind prints falling in front of the fore prints. In soft snow the prints may merge to look like larger pairs of prints; tail drag will be evident. A mouse trail may lead up a tree or down into a burrow.

Similar Species: Many less-common species of mice have near-identical tracks. The House Mouse (*Mus musculus*), with very similar tracks, associates more with humans. Jumping mouse (p. 114) prints show long, thin toes. Voles (p. 110) tend to trot and have a much shorter bounding track pattern. Chipmunks (p. 98) have a wider straddle. Shrews (p. 118) have a narrower straddle.

Meadow Jumping Mouse

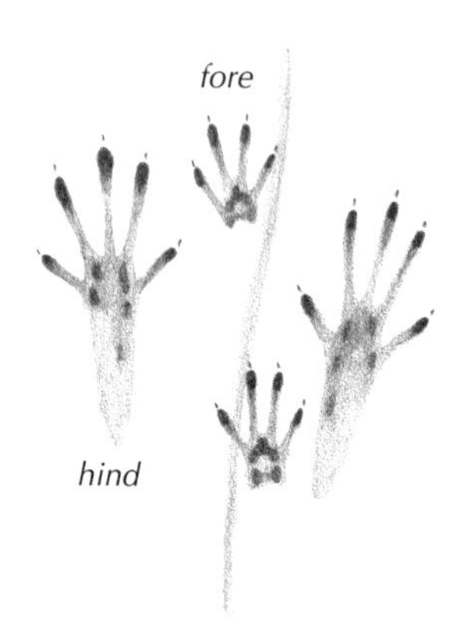

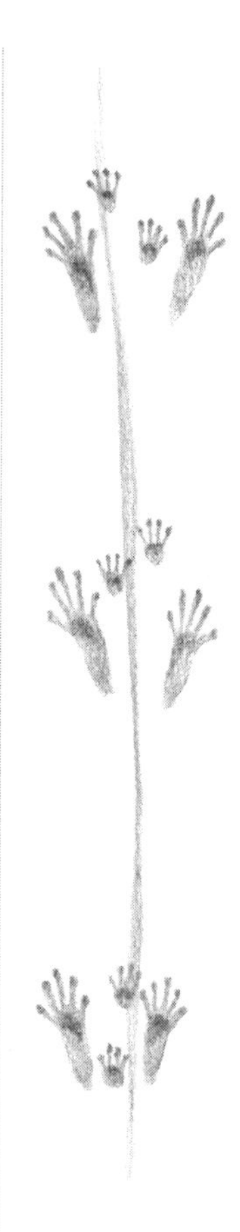
bounding

Fore Print
Length: 0.8–1.3 cm (0.3–0.5 in)
Width: 0.8–1.3 cm (0.3–0.5 in)

Hind Print
Length: 1.3–3.3 cm (0.5–1.3 in)
Width: 1.3–1.8 cm (0.5–0.7 in)

Straddle
4.5–4.8 cm (1.8–1.9 in)

Stride
Bounding: 18–45 cm (7–18 in)
In alarm: 90–180 cm (3–6 ft)

Size
Length with tail: 18–23 cm (7–9 in)

Weight
17–35 g (0.6–1.3 oz)

MEADOW JUMPING MOUSE

Zapus hudsonius

Congratulations if you find and successfully identify the tracks of the Meadow Jumping Mouse! Though it is abundant in northern parts of the province, its preference for grassy meadows and dense undergrowth—and its long, deep winter hibernation (about six months!)—make locating tracks very difficult.

Jumping mouse tracks are distinctive if you do find them. The two smaller forefeet register between the long hind prints; the long heels do not always register, and some prints show just the three long middle toes. The toes on the forefeet may splay so much that the side toes point backward. When they bound, jumping mice make short leaps. The tail may leave a dragline in soft mud or unseasonable snow. Clusters of cut grass stems about 13 cm (5 in) long and lying in meadows are a more abundant sign of this rodent.

Similar Species: Deer Mouse (p. 112) tracks may have the same straddle. Heel-less hind prints may be mistaken for a vole's (p. 110), or a small bird's (p. 140) or even an amphibian's (pp. 142–147).

Ord's Kangaroo Rat

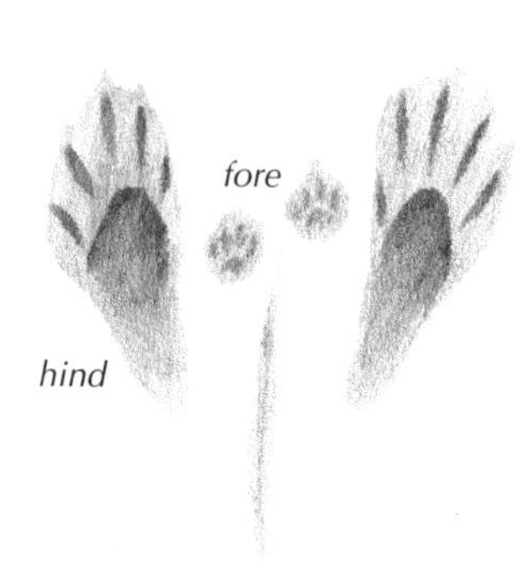

slow hop group

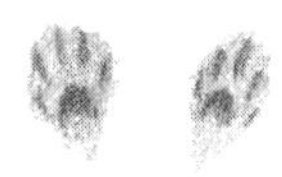

fast hopping

Hind Print
(fore print is much smaller)
Length: 3.8–4.5 cm (1.5–1.8 in)
Width: 1.3–2 cm (0.5–0.8 in)

Straddle
3.3–5.8 cm (1.3–2.3 in)

Stride
Hopping: 13–60 cm (5–24 in)

Size
Length with tail: 20–36 cm (8–14 in)

Weight
43–70 g (1.5–2.5 oz)

ORD'S KANGAROO RAT

Dipodomys ordii

This small, athletic rodent is capable of big jumps, as its name suggests. Sandy, semi-arid parts of southeastern Alberta are the best bet for finding Ord's Kangaroo Rat. In cold weather it stays under the snow, venturing out on milder days.

This nocturnal rodent's preference for drier terrain means that good tracks are hard to find; an abundance of them may be found in sand, but without fine print detail. The best way to identify the track is by its habit. When a kangaroo rat hops slowly, the two small forefeet register between the large hind feet, which show long heel marks, and its long tail leaves a dragline. At higher speed the forefeet do not register, the hind heels appear shorter, and the tail registers infrequently. If you find one of this kangaroo rat's large nesting mounds, tap your fingers beside a burrow—you may hear thumping in reply.

Similar Species: No other kangaroo rats inhabit this region. Other rats and large mice usually register all four feet and their patterns rarely show tail draglines. The Meadow Jumping Mouse (p. 114) prefers lusher areas and leaves much smaller tracks.

Masked Shrew

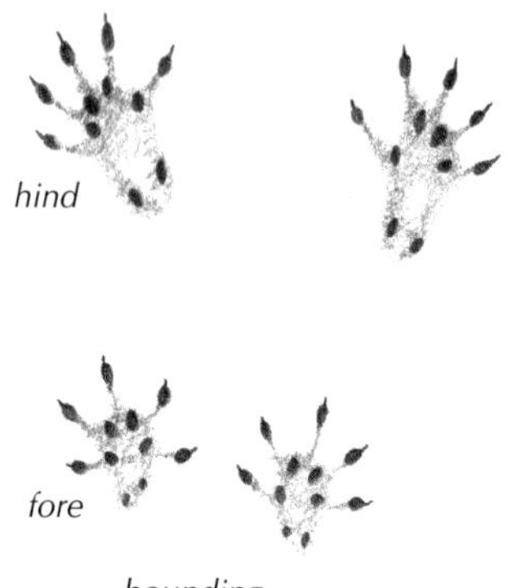

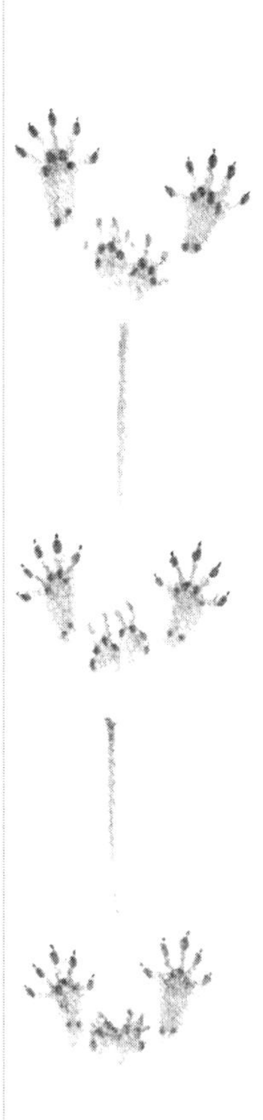

Fore Print
Length: 0.5 cm (0.2 in)
Width: 0.5 cm (0.2 in)

Hind Print
Length: 1.5 cm (0.6 in)
Width: 0.8 cm (0.3 in)

Straddle
2–3.3 cm (0.8–1.3 in)

Stride
Bounding: 3–7.5 cm (1.2–3 in)

Size
Length with tail: 7–11 cm (2.3–4.5 in)

Weight
3–9 g (0.1–0.3 oz)

MASKED SHREW

Sorex cinereus

Though several species of tiny, frenetic shrews are found in Alberta, the widespread and adaptable Masked Shrew is a likely candidate if you find tracks. This small shrew prefers moist fields, marshes, bogs or woodlands, but it can also be found in higher and drier grasslands.

In its energetic and unending quest for food, a shrew usually leaves a four-print bounding pattern, but it may slow to an alternating walking gait. The individual prints in a group are often indistinct, but in mud or shallow, wet snow you can even count the five toes on each print. In deeper snow a shrew's tail often leaves a dragline. If a shrew tunnels under the snow, it may leave a distinct snow ridge on the surface. A shrew's trail may disappear down a burrow.

Similar Species: The Dusky Shrew (*S. monticolus*) and the Arctic Shrew (*S. arcticus*) both make indistinguishable prints. The Pygmy Shrew (*S. hoyi*), also widespread, has slightly smaller prints. A mouse's (pp. 112–115) fore prints will show four toes.

BIRDS, AMPHIBIANS & REPTILES

A guide to the animal tracks of Alberta is not complete without some consideration of the birds, amphibians and reptiles found in the region.

Several bird species have been chosen to represent the main types common to the province, but remember that individual bird species are not easily identified by track alone. Bird tracks are often abundant in snow and are clearest in shallow, wet snow. The shores of lakes and streams are very reliable places to find bird tracks—the mud there can hold a clear print for a long time. The sheer number of tracks made by shorebirds and waterfowl can be astonishing. Though some bird species prefer to perch in trees or soar across the sky, it can be entertaining to follow the tracks of birds that spend a lot of time on the ground. They can spin around in circles and lead you in all directions. The trail may suddenly end as the bird takes flight, or it might terminate in a pile of feathers, the bird having fallen victim to a predator.

Many amphibians and turtles depend on moist environments, so look in the soft mud along the shores of lakes and ponds for their distinctive tracks. You may be able to distinguish frog tracks from toad tracks, because they generally move differently, but it can be very difficult to identify the species. Reptiles thrive and outnumber the amphibians in drier environments, but they seldom leave good tracks, except in occasional mud or perhaps in sand. Snakes leave distinctive body prints.

Canada Goose

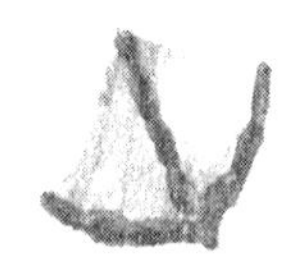

Print
Length: 10–13 cm (4–5 in)

Straddle
13–18 cm (5–7 in)

Stride
Walking: 13–18 cm (5–7 in)

Size
80–120 cm (2.7–4 ft)

CANADA GOOSE

Branta canadensis

This common goose is a familiar sight in open areas by lakes and ponds. Its huge, webbed feet leave prints that can often be seen in abundance along the muddy shores of just about any waterbody, including those in urban parks, where the Canada Goose's green-and-white droppings often accumulate in prolific amounts.

The webbed feet each have three long toes, all facing forward. These toes usually register well, but the webbing between them does not always show in the print. The feet point inward, giving the prints a pigeon-toed appearance and perhaps accounting for the bird's waddling gait.

Similar Species: Many waterfowl, such as ducks (various spp.), as well as the Herring Gull (p. 124), leave similar, usually smaller prints. Exceptionally large prints were likely made by a swan (*Cygnus* spp.).

Herring Gull

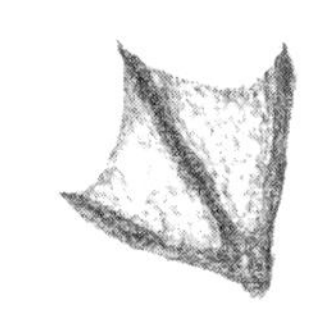

Print
Length: 9 cm (3.5 in)

Straddle
10–15 cm (4–6 in)

Stride
11 cm (4.5 in)

Size
Length: 58–65 cm (23–25 in)

HERRING GULL

Larus argentatus

The Herring Gull, with its long wings and webbed toes, is a strong long-distance flier and an excellent swimmer. Increasingly common in Alberta, it is concentrated in great numbers near waterbodies and garbage dumps.

Gulls leave slightly asymmetrical tracks that show three toes. They have claws that register outside the webbing, and the claw marks are usually attached to the footprint. Most gulls have quite a swagger to their gait, and they leave a trail with the tracks turned strongly inward.

Similar Species: Gull species cannot be reliably identified by track alone, but smaller species have conspicuously smaller tracks. Duck (various spp.) tracks are often difficult to distinguish from gull tracks. Goose (p. 122) tracks are similar but usually larger.

Great Blue Heron

Print
Length: to 17 cm (6.5 in)

Straddle
20 cm (8 in)

Stride
23 cm (9 in)

Size
1.3–1.4 m (4.2–4.5 ft)

GREAT BLUE HERON

Ardea herodias

The regal and graceful image of this large heron has come to symbolize the precious wetlands in which it patiently hunts for food. Usually still and statuesque as it waits for a meal to swim by, this heron will have cause to walk from time to time, perhaps to find a better hunting location. Look for its large, slender tracks along the banks or mudflats of waterbodies.

Not surprisingly, a bird that lives and hunts with such precision walks in a similar fashion, leaving straight tracks that fall in a nearly straight line. Look for the slender rear toe in the print.

Similar Species: The American Bittern (*Botaurus lentiginosus*) makes similar but smaller tracks. Cranes (*Grus* spp.) occupy similar habitats and have similarly sized prints, but a crane's rear toes are smaller and do not register.

Common Snipe

Print
Length: 3.8 cm (1.5 in)

Straddle
to 4.5 cm (1.8 in)

Stride
to 3.3 cm (1.3 in)

Size
28–30 cm (11–12 in)

COMMON SNIPE

Gallinago gallinago

This short-legged character is a resident of marshes and bogs, where its neat prints can often be seen in mud. Snipes are quite secretive when on the ground, so you may be surprised if one suddenly flushes out from beneath your feet. If there is a Common Snipe in the air, you may hear an eerie whistle as it dives from the sky.

The Common Snipe's neat prints show four toes, including a small rear toe that points inward. The bird's short legs and stocky body give it a very short stride.

Similar Species: Many shorebirds, including the Spotted Sandpiper (p. 130), leave similar tracks.

Spotted Sandpiper

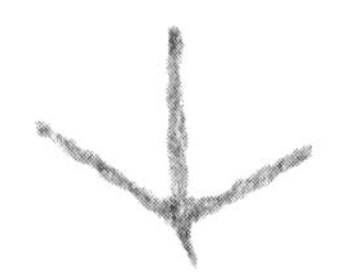

Print
Length: 2–3.3 cm (0.8–1.3 in)

Straddle
to 3.8 cm (1.5 in)

Stride
Erratic

Size
18–20 cm (7–8 in)

SPOTTED SANDPIPER

Actitis macularia

The bobbing tail of the Spotted Sandpiper is a common sight on the shores of lakes, rivers and streams, but you will usually find just one of these territorial birds in any given location. Because of its excellent camouflage, likely the first that you will see of this bird will be when it flies away, its fluttering wings close to the surface of the water.

As it teeters up and down on the shore, a sandpiper leaves trails of three-toed prints. Its fourth toe is very small and faces off to one side at an angle. Sandpiper tracks often have an erratic stride.

Similar Species: All sandpipers and plovers, including the common Killdeer (*Charadrius vociferus*), leave similar tracks, although there is much diversity in size. The Common Snipe (p. 128) makes similar but larger tracks.

Ruffed Grouse

Print
Length: 5–7.5 cm (2–3 in)

Straddle
5–7.5 cm (2–3 in)

Stride
Walking: 7.5–15 cm (3–6 in)

Size
38–48 cm (15–19 in)

RUFFED GROUSE

Bonasa umbellus

This ground-dweller prefers the quiet seclusion of coniferous forests, where its excellent camouflage usually affords it good protection. Although you might find its tracks in mud, snow much improves your chances of finding them. If you follow a Ruffed Grouse trail quietly, you may be startled when the bird bursts from cover almost beneath your feet.

The three thick front toes leave very clear impressions, but the short rear toe, which is angled off to one side, does not always show up so well. This bird's neat, straight trail appears to reflect its conservative and cautious approach to life on the forest floor.

Similar Species: Other grouse and ptarmigans (*Lagopus* spp.) leave similar tracks, but their prints may be blurred and enlarged by the winter feathers that grow on their feet.

Great Horned Owl

Strike
Width: to 90 cm (3 ft)

Size
55 cm (22 in)

GREAT HORNED OWL

Bubo virginianus

Often seen resting quietly in trees by day, this wide-ranging owl prefers to hunt at night. You might find an untidy hole in the snow, possibly surrounded by wing and tail-feather imprints—a well-registered 'strike' can be quite a sight. The Great Horned Owl strikes through the snow with its talons, and the feather imprints are made as the owl struggles to take off with possibly heavy prey. An ungraceful walker, it prefers to fly away from the scene, though its tracks may be evident near roadkill.

You might stumble across a strike and guess that the owl's target could have been a vole (p. 110) scurrying around underneath the snow. Or you may be following the surface trail of an animal, only to find that it abruptly ends with a strike mark where the animal has been seized.

Similar Species: If the prey left no approaching trail, the strike is likely an owl's, because owls hunt by sound. Otherwise, the strike—usually with more distinct feather imprints—could be that of a hawk or a Common Raven (*Corvus corax*), both of which hunt by sight.

Burrowing Owl

Print
Length: 4.5 cm (1.8 in)

Straddle
6.5 cm (2.5 in)

Stride
2.5–15 cm (1–6 in)

Size
24 cm (9.5 in)

BURROWING OWL

Athene cunicularia

This alert little owl of the plains and open areas spends a lot of time on the ground, bobbing up and down to look out for danger. It inhabits abandoned burrows, such as those made by ground squirrels (p. 96), often close to burrows still inhabited by the rodents. Not a year-round resident, the Burrowing Owl migrates south for the winter.

If you are investigating rodent burrows in open areas and you find a burrow entrance with a profusion of tracks that do not have characteristic rodent features, you may have found the unusual tracks of the Burrowing Owl. The print of this owl shows two large, forward-pointing toes with talons, a shorter toe to the side and a very short fourth one to the rear. The toes to the side and rear do not register as well as the front toes, which have more weight on them.

Similar Species: The tracks of other owls are similar, but are unlikely to be found in Burrowing Owl habitat.

American Crow

Print
Length: 6.5–7.5 cm (2.5–3 in)

Straddle
3.8–7.5 cm (1.5–3 in)

Stride
Walking: 10 cm (4 in)

Size
40 cm (16 in)

AMERICAN CROW

Corvus brachyrhyncos

The black silhouette of the American Crow can be a common sight in a variety of habitats. A crow will frequently come down to the ground and contentedly strut around. Its loud *caw* can be heard from quite a distance. Crows can be especially noisy when they are mobbing an owl or a hawk.

The American Crow typically leaves an alternating walking track pattern. Its prints show three sturdy toes pointing forward and one toe pointing backward. When a crow is in need of greater speed, perhaps for take-off, it bounds along, leaving irregular pairs of diagonally placed prints with a longer stride between each pair.

Similar Species: Other corvids also spend a lot of time on the ground and make similar tracks. The Black-billed Magpie (*Pica pica*) makes prints up to 5 cm (2 in) long. The much larger Common Raven (*Corvus corax*) leaves tracks to 10 cm (4 in) long, with a stride of 15 cm (6 in).

Dark-eyed Junco

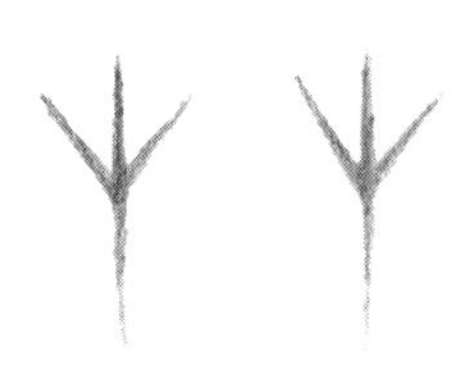

Print
Length: to 3.8 cm (1.5 in)
Straddle
2.5–3.8 cm (1–1.5 in)
Stride
Hopping: 3.8–13 cm (1.5–5 in)
Size
14–17 cm (5.5–6.5 in)

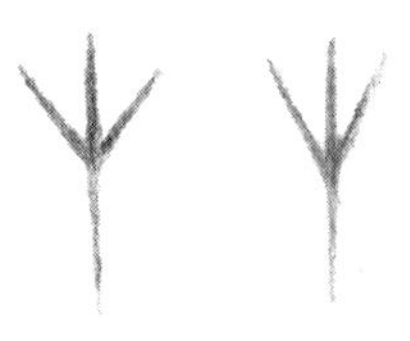

DARK-EYED JUNCO

Junco hyemalis

This common small bird typifies the many small hopping birds found in the province. Each foot has three forward-pointing toes and one longer toe at the rear. The best prints are left in snow, although in deep snow the toe detail is lost; the footprints may show some dragging between the hops.

A good place to study this type of prints is near a birdfeeder. Watch the birds scurry around as they pick up fallen seeds, then have a look at the prints left behind. For example, juncos are attracted to small seeds that chickadees (*Poecile* spp.) scatter as they forage for sunflower seeds in the birdfeeder. Also look for tracks under coniferous trees, where juncos feed on fallen seeds in winter.

Similar Species: Note that the changing seasons influence the bird diversity in a region. Toe size may help with identification, because larger birds make larger prints. In powdery snow, junco tracks could be confused with mouse (pp. 112–115) tracks, so follow the trail to see if it disappears down a hole or into thin air.

Frogs

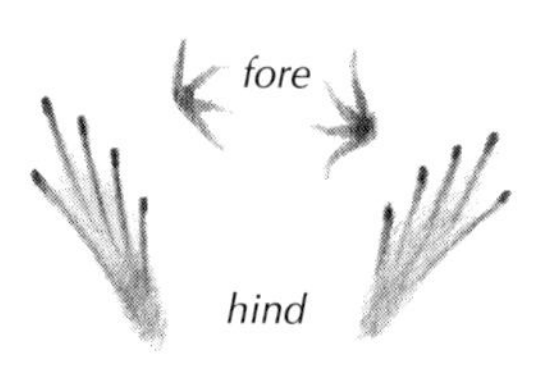

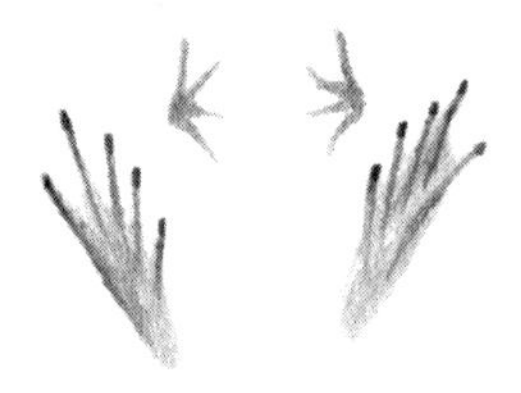

Straddle
to 7.5 cm (3 in)

hopping

FROGS

The best place to look for frog tracks is along the muddy fringes of waterbodies.

Among the frogs whose tracks you might find is the small Boreal Chorus Frog (*Pseudacris maculata*), which is common throughout the province. The larger Wood Frog (*Rana sylvatica*) is found throughout the province as well, but especially in waterbodies bordering woodlands. The much larger Northern Leopard Frog (*R. pipiens*) prefers marshes and moist fields, wherever there is profuse grassy vegetation.

A frog's hopping action results in its two small forefeet registering in front of its long-toed hind prints. Frog tracks vary greatly in size, depending on species and age. Toads (p. 144) may also hop, but they usually walk.

Toads

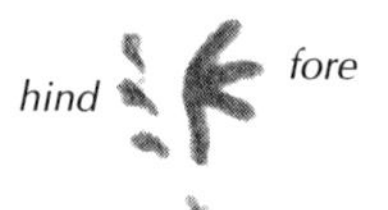

Straddle
to 6.5 cm (2.5 in)

walking

TOADS

Plains Spadefoot

The best place to look for toad tracks is, as with frogs, undoubtedly along the muddy fringes of waterbodies, but their tracks can also be found in drier areas as unclear trails in dusty patches of soil.

One of the most widespread toads in the province is the Canada Toad (*Bufo hemiophrys*), which can be found near quiet streams and sometimes in woodland areas. The Plains Spadefoot (*Spea bombifrons*) inhabits shortgrass prairie in the southernmost part of the province. The Boreal Toad (*B. boreas*), found in the mountain and central regions, is frequently seen on moist forest floors. The elusive Great Plains Toad (*B. cognatus*), found only in the southeastern corner of Alberta, is the only other toad likely to be encountered.

In general, toads walk and frogs (p. 142) hop, but toads are pretty capable hoppers, too, especially when they are hassled by overly enthusiastic naturalists. Toads leave rather abstract prints as they walk. The heels of the hind feet do not register. On less-firm surfaces, the toes often leave draglines.

Salamanders & Lizards

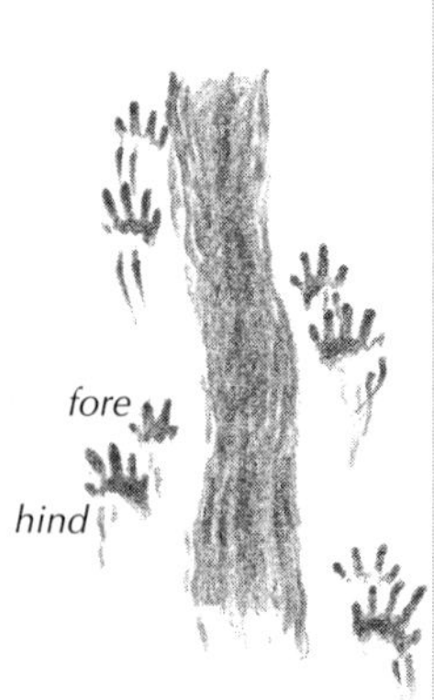

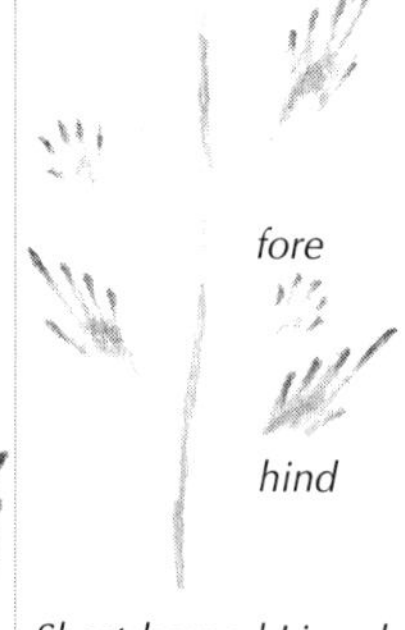

Straddle
to 7.5 cm (3 in)

salamander walking

Short-horned Lizard walking

SALAMANDERS & LIZARDS

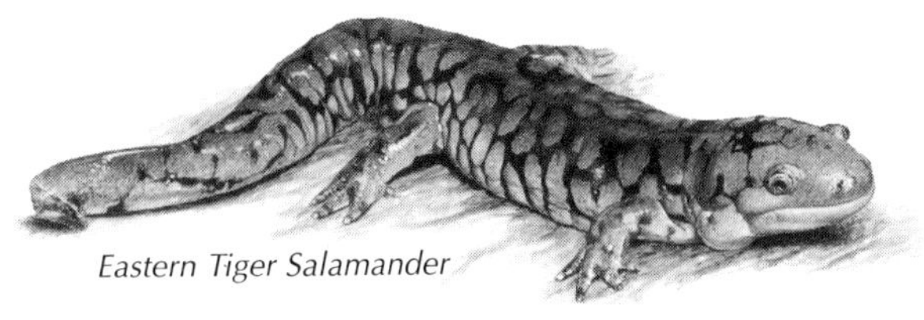

Eastern Tiger Salamander

The harsh seasons of Alberta are ill-suited for most species of salamanders and lizards. Therefore, there are only a few of these species in the province.

Salamanders—long, slender, lizard-like amphibians—need a relatively moist environment. The Eastern Tiger Salamander (*Ambystoma tigrinum*) can be found throughout the southern part of the province on sagebrush plains, in open forests and in damp meadows. The Long-toed Salamander (*A. macrodactylum*) lives only in the mountains.

In southeastern Alberta, the Short-horned Lizard (*Phrynosoma douglasii*) might be found sunning itself on sandy plains or on rocky outcrops in forested areas. This unusual flat-bodied lizard grows to about 15 cm (6 in) long.

Because grass and rocks do not retain track impressions, finding salamander or lizard tracks is difficult. Worse, print detail is often blurred by the animal's dragging belly or by the swinging of its thick tail across its tracks. If you do find clear tracks, perhaps in dust or mud, four-toed fore prints and shorter toe marks will help you to distinguish salamander tracks from otherwise similar lizard tracks.

Turtles

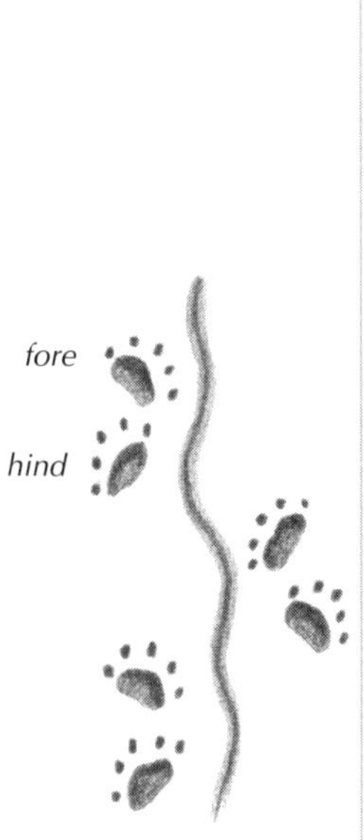

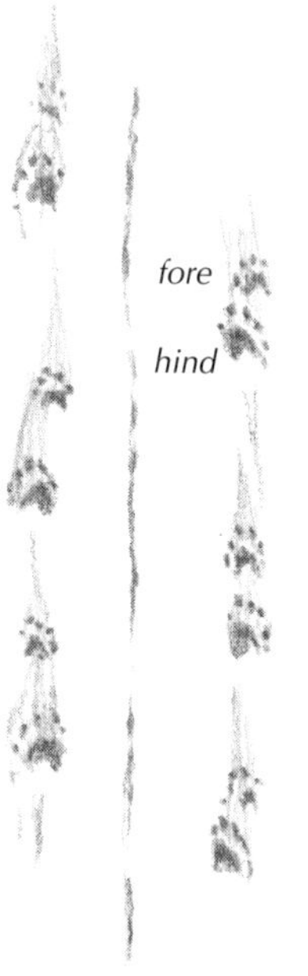

Straddle
10–25 cm (4–10 in)

Snapping Turtle walking

Painted Turtle walking

TURTLES

Most turtles enjoy watery environs and warm seasons, so they are not plentiful in this province. Nevertheless, along the muddy edges of slow-moving streams, marshes or ponds in Alberta's far southeasternmost corner you might find the tracks of the Snapping Turtle (*Chelydra serpentina*), which grows to 48 cm (19 inches) long. The Snapping Turtle is highly aquatic and rarely ventures out of the water, so its tracks are not common.

The much smaller Painted Turtle (*Chrysemys picta*), which grows to only 25 cm (10 in) in length, has a range that extends across much of extreme southern Alberta. This pretty turtle is found in slow-moving and shallow streams, rivers and lakes.

Because of its large shell and short legs, a turtle's trail is wide relative to the length of its stride, and its straddle is about half its body length. On firmer surfaces, look for distinct claw marks. Although longer-legged turtles can raise their shells off the ground, short-legged species may let their shells drag, as shown in their tracks. The tail may leave a straight dragline in the mud.

Snakes

SNAKES

Common Garter Snake

Of the many snakes that inhabit Alberta, the Common Garter Snake (*Thamnophis sirtalis*) is the most frequently encountered. Found throughout much of the province, often close to wet or moist areas, this harmless snake, one of three garters in the province, can grow to 1.3 m (4.3 ft) long.

The Western Rattlesnake (*Crotalus viridis*) is the only rattlesnake to be encountered in the province. Found in the south, but not in the far west, it reaches a length of 1.6 m (5.3 ft) and frequents a variety of habitats that range from marshlands to dry woodlands. The larger Bullsnake (*Pituophis catenifer*) has a similar range. The smaller Western Hognose Snake (*Heterodon nasicus*) lives in sandy, open areas of the far southeast.

Identifying a snake by its tracks is next to impossible. It is very challenging even to establish which way the snake was moving, but a wide trail with strong side looping indicates that the snake was moving quickly, whereas narrow and uneven trails result from lower speeds.

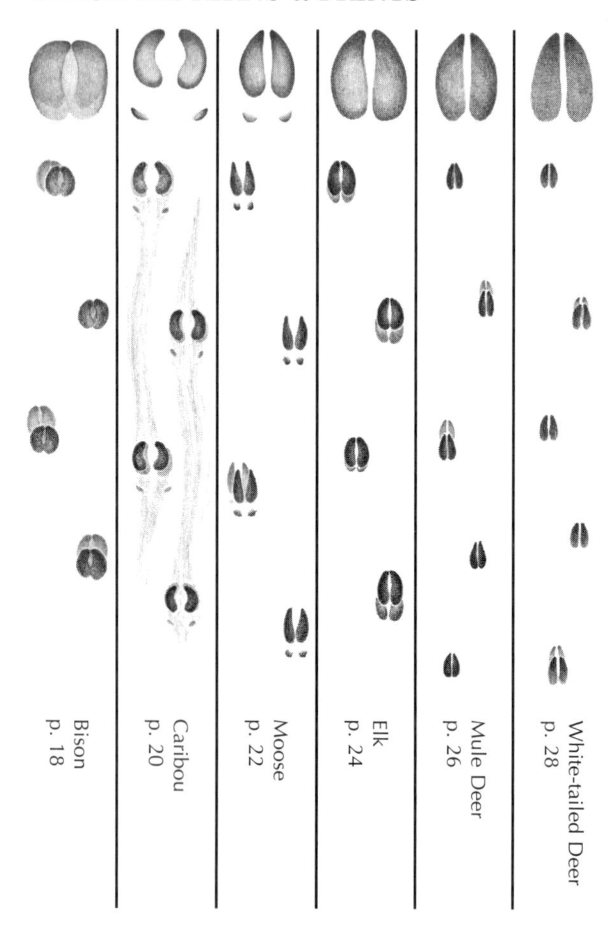
Bison
p. 18
Caribou
p. 20
Moose
p. 22
Elk
p. 24
Mule Deer
p. 26
White-tailed Deer
p. 28

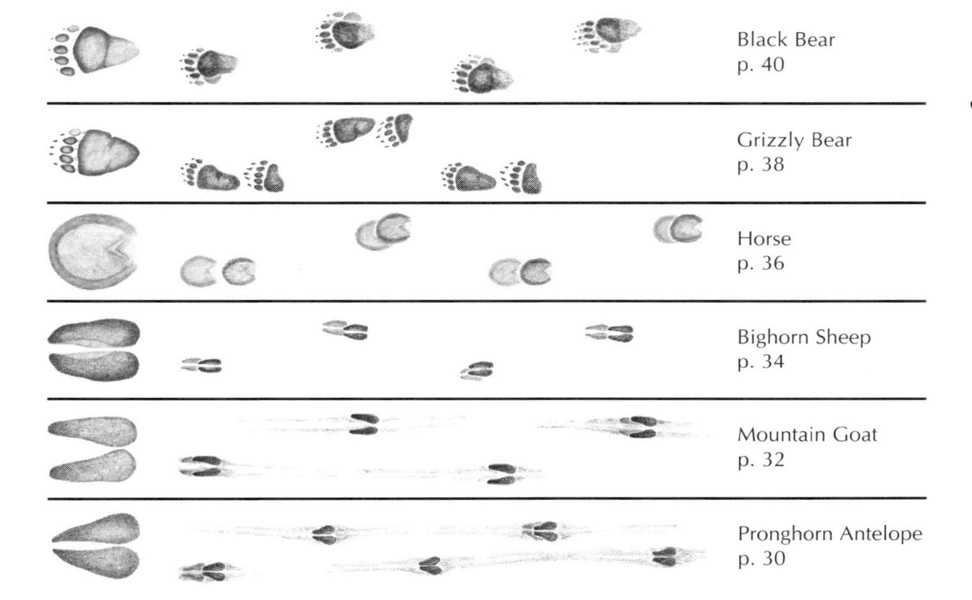
Black Bear
p. 40
Grizzly Bear
p. 38
Horse
p. 36
Bighorn Sheep
p. 34
Mountain Goat
p. 32
Pronghorn Antelope
p. 30

TRACK PATTERNS & PRINTS

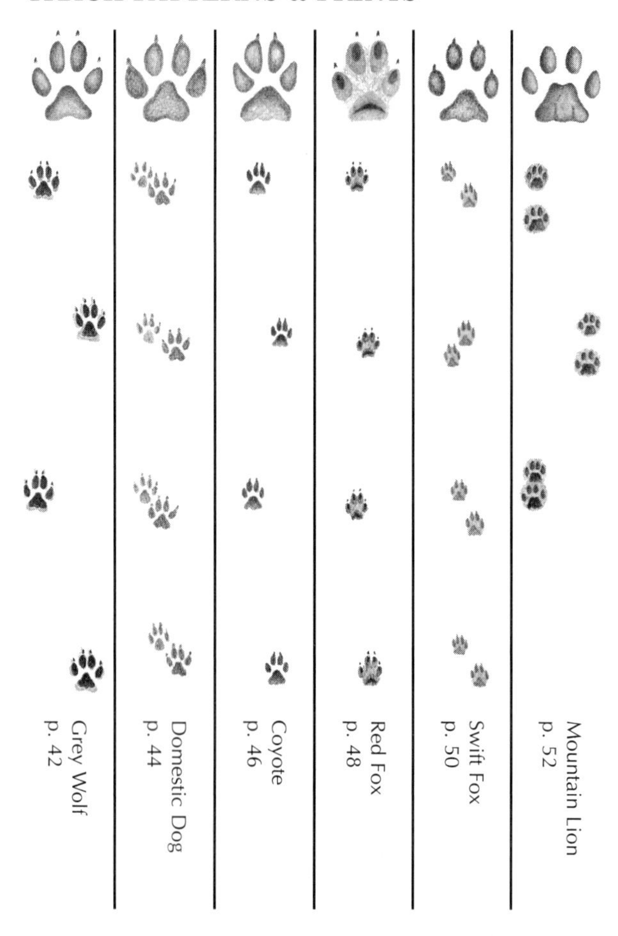

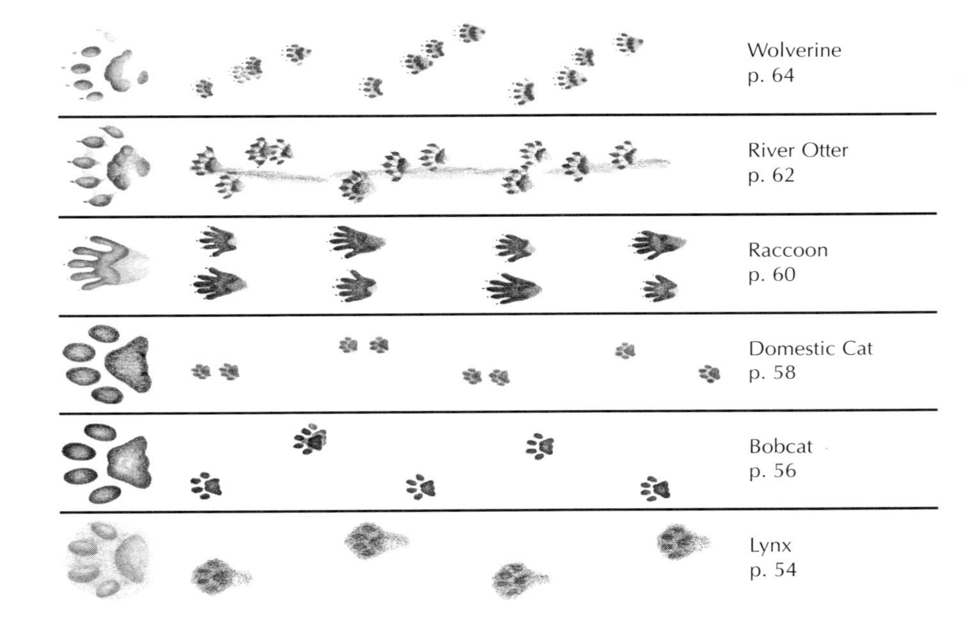
Wolverine
p. 64
River Otter
p. 62
Raccoon
p. 60
Domestic Cat
p. 58
Bobcat
p. 56
Lynx
p. 54

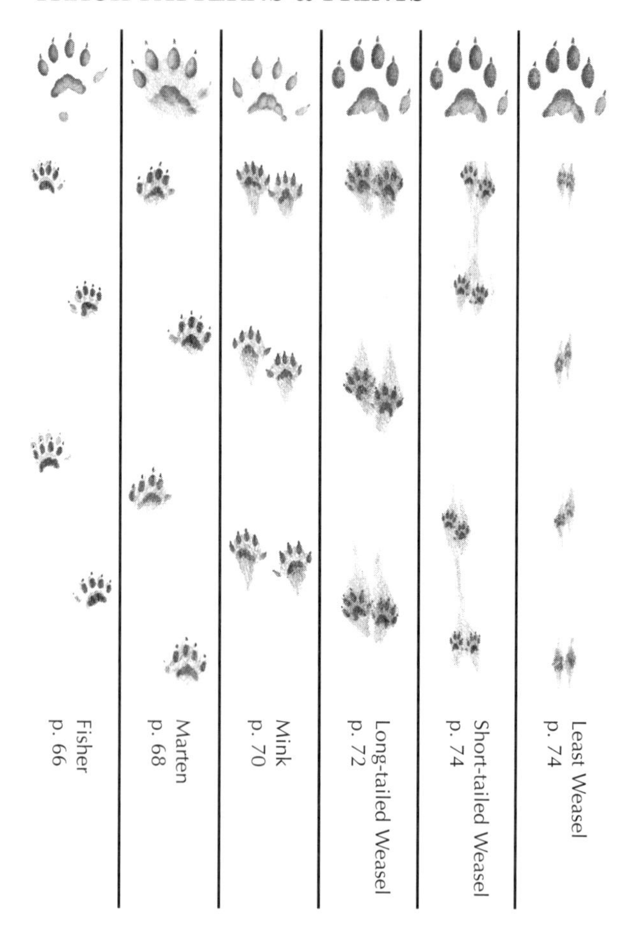
Fisher
p. 66
Marten
p. 68
Mink
p. 70
Long-tailed Weasel
p. 72
Short-tailed Weasel
p. 74
Least Weasel
p. 74

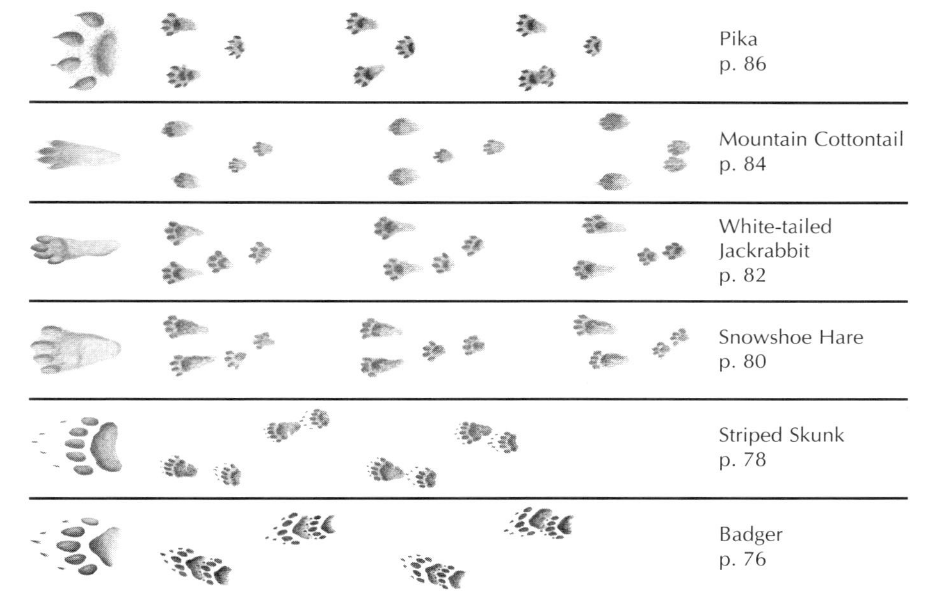

Pika
p. 86
Mountain Cottontail
p. 84
White-tailed
Jackrabbit
p. 82
Snowshoe Hare
p. 80
Striped Skunk
p. 78
Badger
p. 76

TRACK PATTERNS & PRINTS

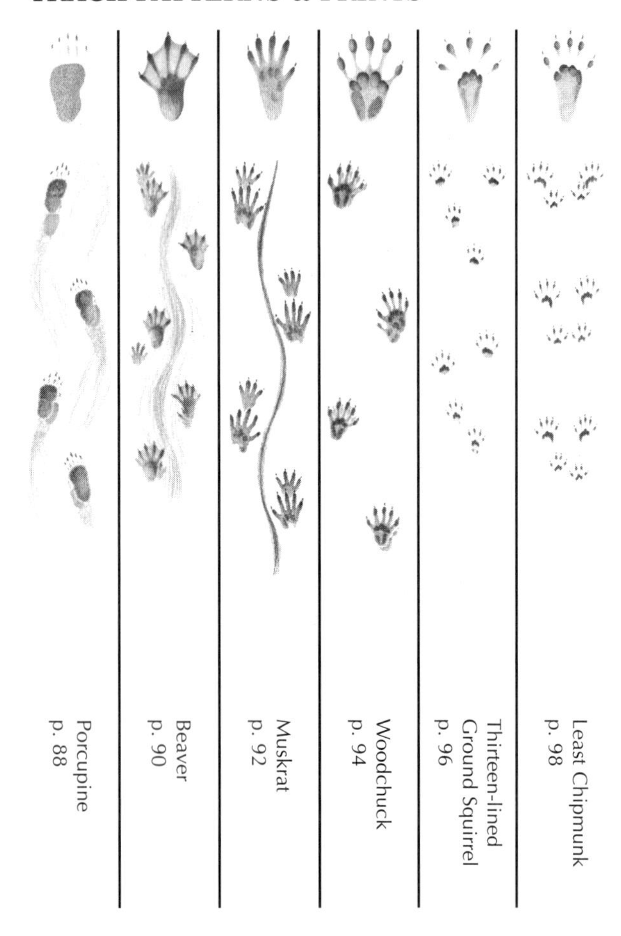

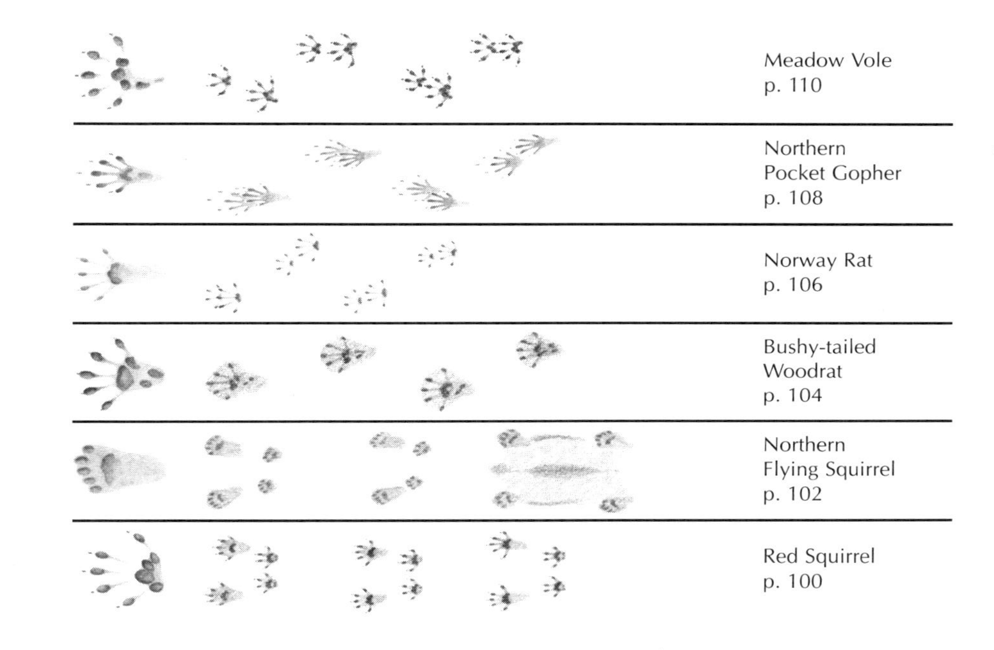

Meadow Vole
p. 110
Northern
Pocket Gopher
p. 108
Norway Rat
p. 106
Bushy-tailed
Woodrat
p. 104
Northern
Flying Squirrel
p. 102
Red Squirrel
p. 100

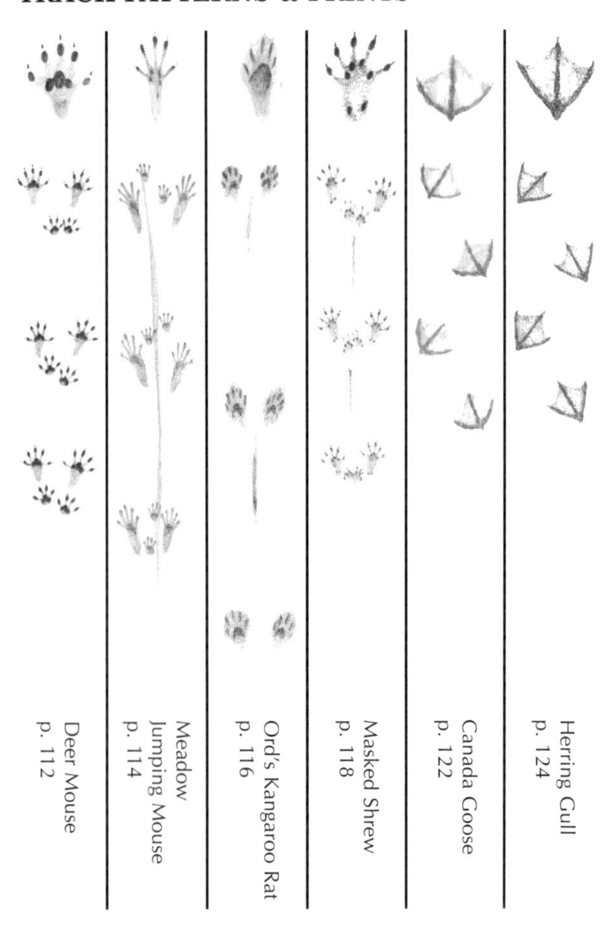
Deer Mouse
p. 112
Meadow
Jumping Mouse
p. 114
Ord's Kangaroo Rat
p. 116
Masked Shrew
p. 118
Canada Goose
p. 122
Herring Gull
p. 124

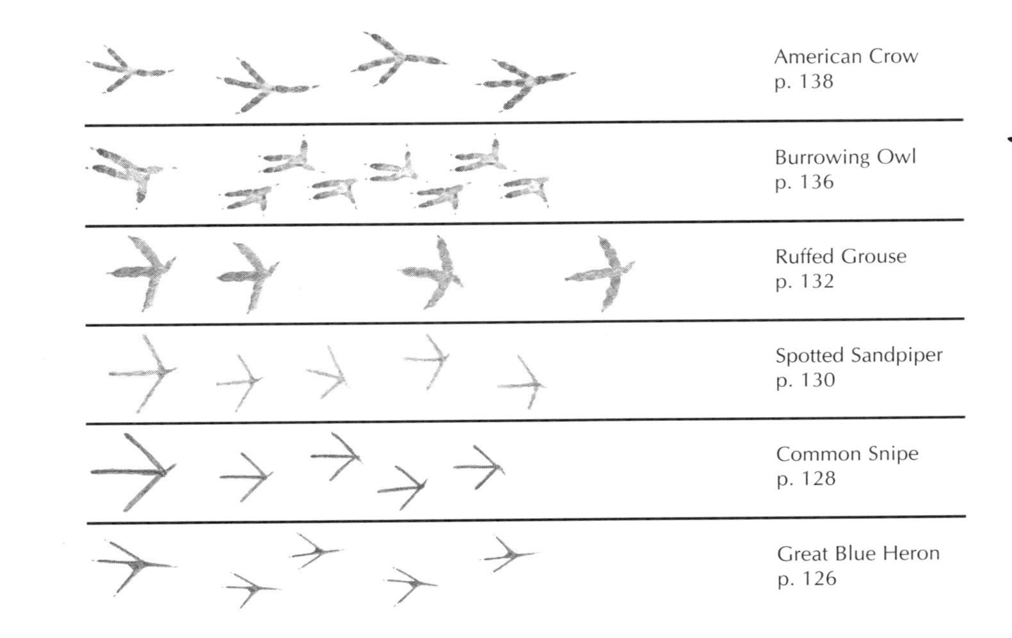
American Crow
p. 138
Burrowing Owl
p. 136
Ruffed Grouse
p. 132
Spotted Sandpiper
p. 130
Common Snipe
p. 128
Great Blue Heron
p. 126

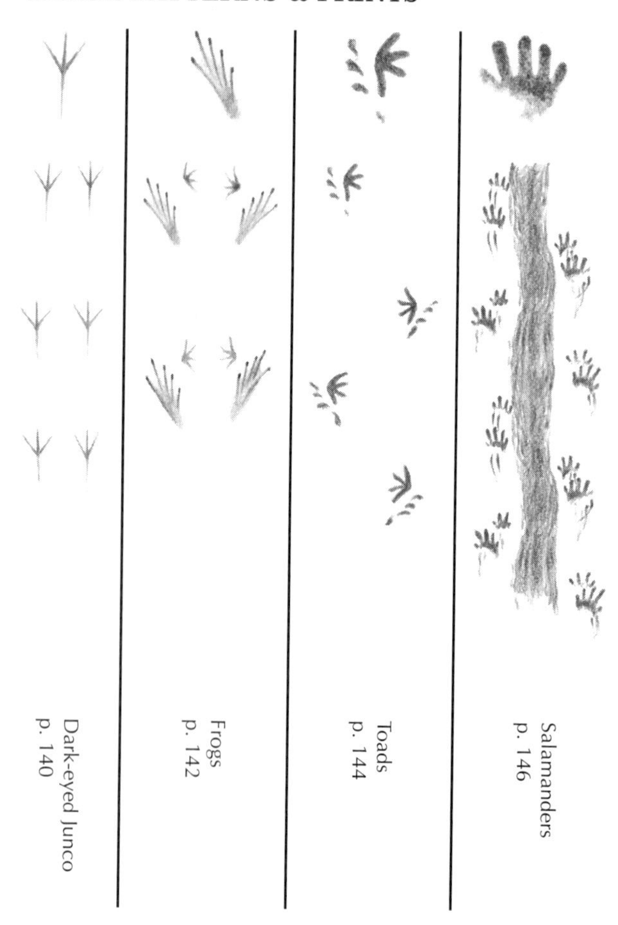
Dark-eyed Junco
p. 140
Frogs
p. 142
Toads
p. 144
Salamanders
p. 146

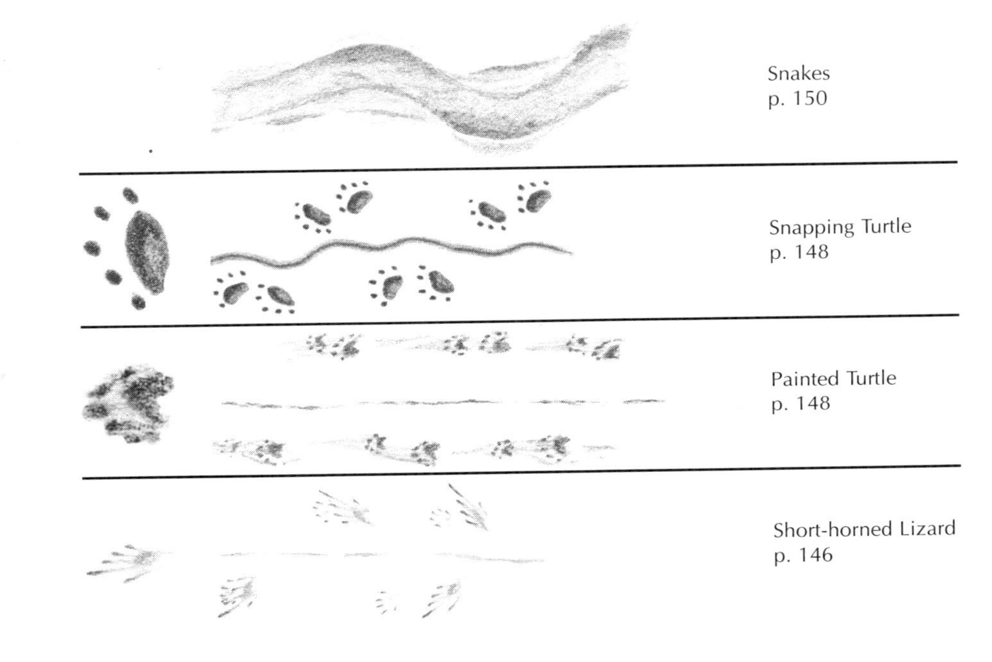
Snakes
p. 150
Snapping Turtle
p. 148
Painted Turtle
p. 148
Short-horned Lizard
p. 146

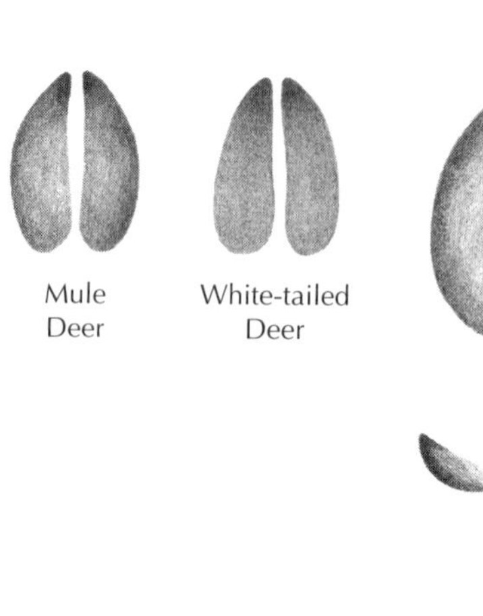

Mule Deer

White-tailed Deer

Caribou

Elk

inch cm
0 0
1
2 5

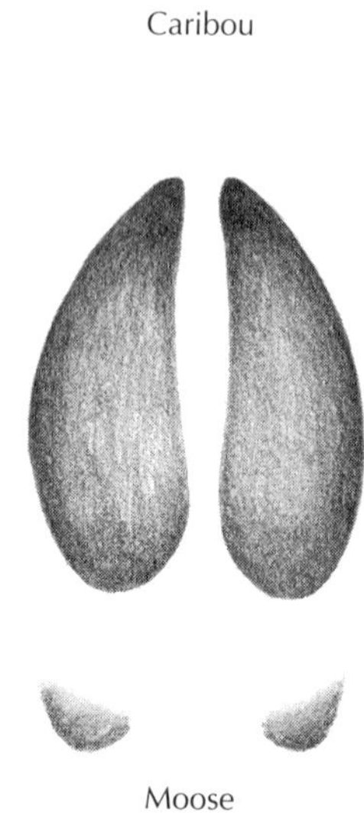

Moose

HOOFED PRINTS

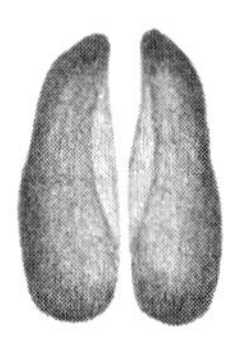

Bighorn
Sheep

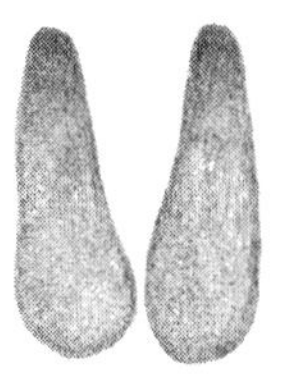

Mountain
Goat

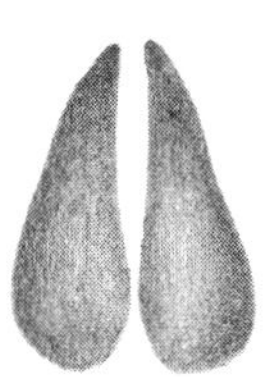

Pronghorn
Antelope

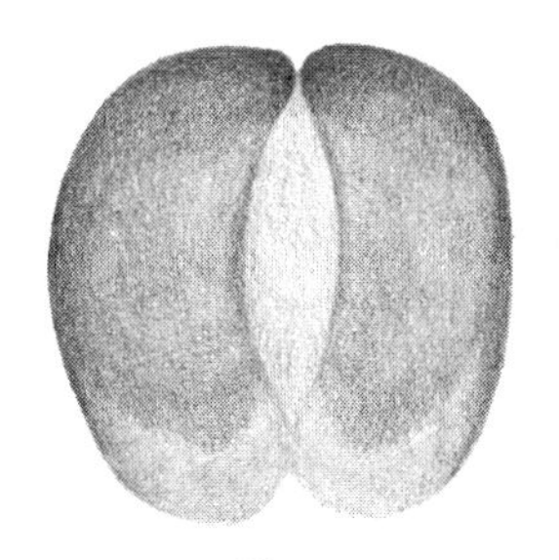

Bison

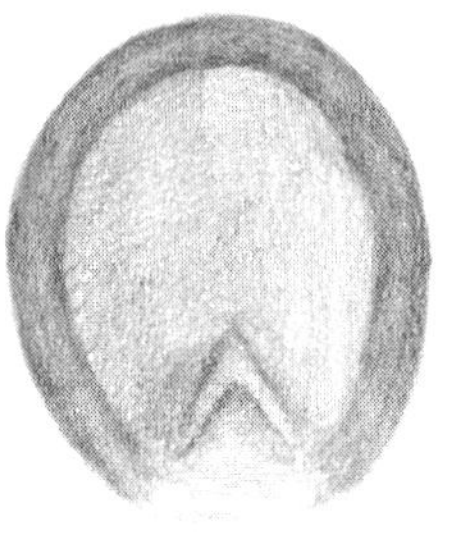

Horse

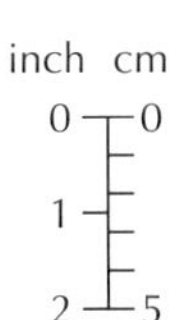

FORE PRINTS

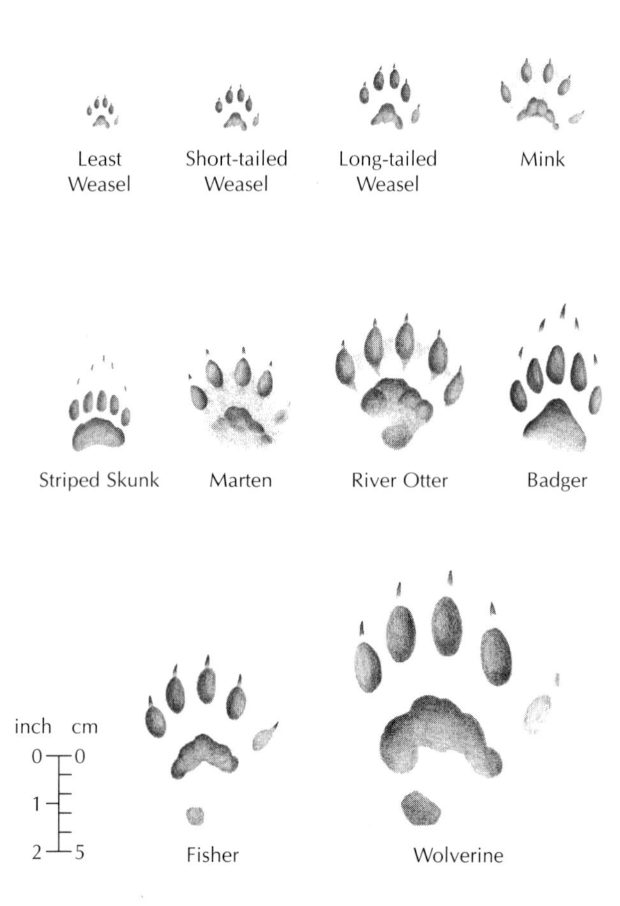
Least Weasel
Short-tailed Weasel
Long-tailed Weasel
Mink
Striped Skunk
Marten
River Otter
Badger
inch cm
0 0
1
2 5
Fisher
Wolverine

FORE PRINTS

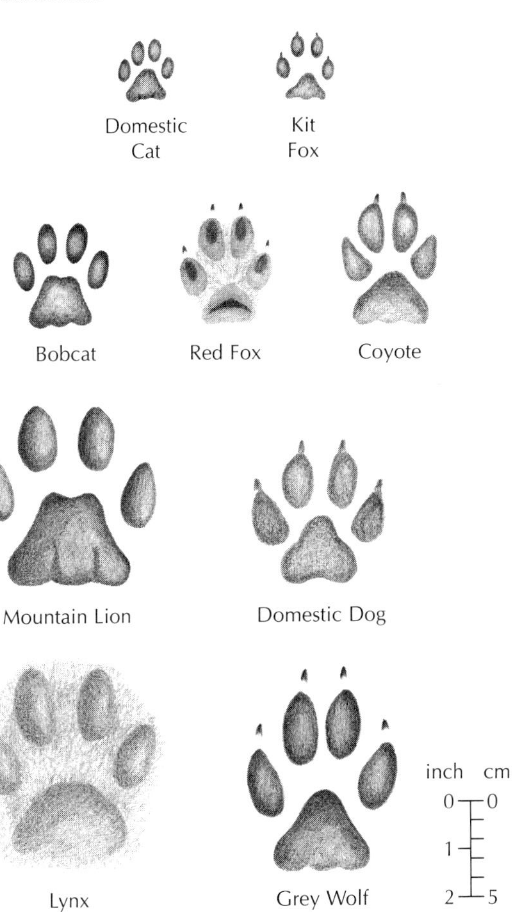
Domestic
Cat
Kit
Fox
Bobcat
Red Fox
Coyote
Mountain Lion
Domestic Dog
Lynx
Grey Wolf
inch cm
0 0
1
2 5

HIND PRINTS

Pika

Norway Rat

Bushy-tailed Woodrat

Thirteen-lined Ground Squirrel

Ord's Kangaroo Rat

Northern Flying Squirrel

Red Squirrel

Woodchuck

Muskrat

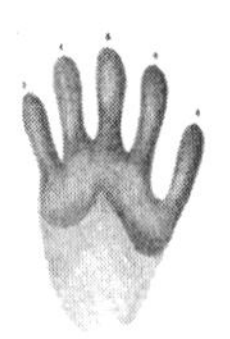
Raccoon

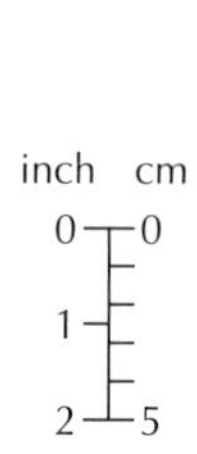

Porcupine

Mountain Cottontail

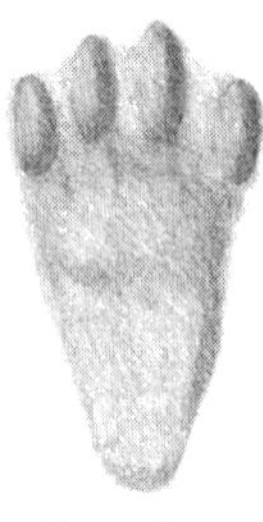
Snowshoe Hare

White-tailed Jackrabbit

HIND PRINTS

Deer
Mouse

Masked
Shrew

Meadow
Vole

Northern
Pocket
Gopher

Meadow
Jumping Mouse

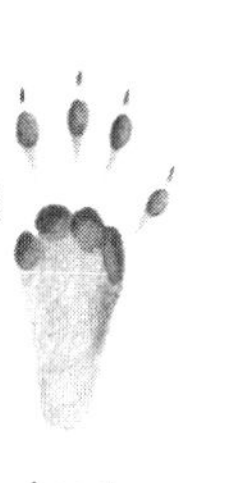

Least
Chipmunk

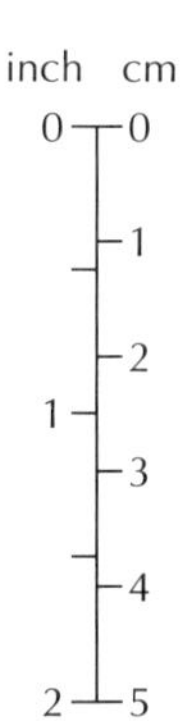

HIND PRINTS

Beaver

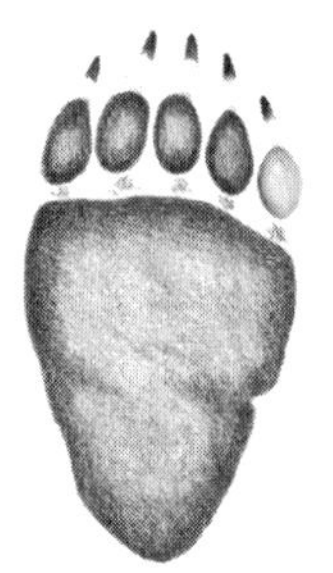

Black Bear

BIBLIOGRAPHY

Behler, J.L., and F.W. King. 1979. *Field Guide to North American Reptiles and Amphibians.* National Audubon Society. New York: Alfred A. Knopf.

Brown, R., J. Ferguson, M. Lawrence and D. Lees. 1987. *Tracks and Signs of the Birds of Britain and Europe: An Identification Guide.* London: Christopher Helm.

Burt, W.H. 1976. *A Field Guide to the Mammals.* Boston: Houghton Mifflin Company.

Farrand, J., Jr. 1995. *Familiar Animal Tracks of North America.* National Audubon Society Pocket Guide. New York: Alfred A. Knopf.

Forrest, L.R. 1988. *Field Guide to Tracking Animals in Snow.* Harrisburg: Stackpole Books.

Halfpenny, J. 1986. *A Field Guide to Mammal Tracking in North America.* Boulder: Johnson Publishing Company.

Headstrom, R. 1971. *Identifying Animal Tracks.* Toronto: General Publishing Company.

Murie, O.J. 1974. *A Field Guide to Animal Tracks.* The Peterson Field Guide Series. Boston: Houghton Mifflin Company.

Rezendes, P. 1992. *Tracking and the Art of Seeing: How to Read Animal Tracks and Sign.* Vermont: Camden House Publishing.

Stall, C. 1989. *Animal Tracks of the Rocky Mountains.* Seattle: The Mountaineers.

Stokes, D., and L. Stokes. 1986. *A Guide to Animal Tracking and Behaviour.* Toronto: Little, Brown and Company.

Wassink, J.L. 1993. *Mammals of the Central Rockies.* Missoula: Mountain Press Publishing Company.

Whitaker, J.O., Jr. 1996. *National Audubon Society Field Guide to North American Mammals.* New York: Alfred A. Knopf.

INDEX

ABOUT THE AUTHORS

Ian Sheldon, an accomplished artist, naturalist and educator, has lived in South Africa, Singapore, Britain and Canada. Caught collecting caterpillars at the age of three, he has been exposed to the beauty and diversity of nature ever since. He was educated at Cambridge University and the University of Alberta. When he is not in the tropics working on conservation projects or immersing himself in our beautiful wilderness, he is sharing his love for nature. Ian enjoys communicating this passion through the visual arts and the written word.

Tamara Eder, equipped from the age of six with a canoe, a dip net and a note pad, grew up with a fascination for nature and the diversity of life. She has a degree in environmental conservation sciences and has photographed and written about the biodiversity in Bermuda, the Galapagos Islands, the Amazon Basin, China, Tibet, Vietnam, Thailand and Malaysia.